# Future Directions for the National Science Foundation's Arctic Natural Sciences Program

Polar Research Board
Commission on Geosciences, Environment, and Resources
National Research Council

NATIONAL ACADEMY PRESS
Washington, D.C. 1998

**NOTICE**: The project that is the subject of this report was approved by the Governing Board of the National Research Council, whose members are drawn from the councils of the National Academy of Sciences, the National Academy of Engineering, and the Institute of Medicine. The members of the committee responsible for writing this report were chosen for their special competencies and with regard for appropriate balance.

Support for this project was provided by the National Science Foundation under Contract OPP-9707783.

Limited copies of this report are available from:

Polar Research Board
National Research Council
2101 Constitution Avenue, NW
Washington, DC 20418
202-334-3479

or

National Academy Press
2101 Constitution Avenue, NW
Lockbox 285
Washington, DC 20055
202-334-3313
800-624-6242
http://www.nap.edu/bookstore

*Cover:* Researchers investigating the origin of Cretaceous flood basalt volcanism in the Arctic (Axel Heiberg Island) under a grant provided by the Arctic Natural Sciences Program. Flood basalts are exposed on islands throughout the High Arctic and may comprise major bathymetric features of the Arctic Ocean such as Alpha Ridge. Photo provided by principal investigator Dr. John A. Tarduno, Department of Earth and Environmental Sciences, University of Rochester, New York. Other images can be viewed at http://www.earth.rochester.edu/pmag/arctic.



Printed in the United States of America

## COMMITTEE ON SCIENCE PRIORITIES FOR NSF'S ARCTIC NATURAL SCIENCES PROGRAM

JOHN T. ANDREWS, *Chair*, University of Colorado, Boulder
SUSAN K. AVERY, University of Colorado, Boulder
MARIANNE S.V. DOUGLAS, University of Toronto, Ontario
BERNARD HALLET, University of Washington, Seattle
PAUL A. MAYEWSKI, University of New Hampshire, Durham
JAMES H. MORISON, University of Washington, Seattle
KIM M. PETERSON, University of Alaska, Anchorage
DONALD B. SINIFF, University of Minnesota, St. Paul
ROGER W. SMITH, University of Alaska, Fairbanks

*Staff*

CHRIS ELFRING, Director
JUDY ESTEP, Administrative Assistant
ROB GREENWAY, Project Assistant

## POLAR RESEARCH BOARD

The National Academy of Sciences is a private, nonprofit, self-perpetuating society of distinguished scholars engaged in scientific and engineering research, dedicated to the furtherance of science and technology and to their use for the general welfare. Upon the authority of the charter granted to it by the Congress in 1863, the Academy has a mandate that requires it to advise the federal government on scientific and technical matters. Dr. Bruce M. Alberts is president of the National Academy of Sciences.

The National Academy of Engineering was established in 1964, under the charter of the National Academy of Sciences, as a parallel organization of outstanding engineers. It is autonomous in its administration and in the selection of its members, sharing with the National Academy of Sciences the responsibility for advising the federal government. The National Academy of Engineering also sponsors engineering programs aimed at meeting national needs, encourages education and research, and recognizes the superior achievements of engineers. Dr. William A. Wulf is president of the National Academy of Engineering.

The Institute of Medicine was established in 1970 by the National Academy of Sciences to secure the services of eminent members of appropriate professions in the examination of policy matters pertaining to the health of the public. The Institute acts under the responsibility given to the National Academy of Sciences by its congressional charter to be an adviser to the federal government and, upon its own initiative, to identify issues of medical care, research, and education. Dr. Kenneth I. Shine is president of the Institute of Medicine.

The National Research Council was organized by the National Academy of Sciences in 1916 to associate the broad community of science and technology with the Academy's purposes of furthering knowledge and advising the federal government. Functioning in accordance with general policies determined by the Academy, the Council has become the principal operating agency of both the National Academy of Sciences and the National Academy of Engineering in providing services to the government, the public, and the scientific and engineering communities. The Council is administered jointly by both Academies and the Institute of Medicine. Dr. Bruce M. Alberts and Dr. William A. Wulf are chairman and vice chairman, respectively, of the National Research Council.

# Preface

The National Science Foundation (NSF) has a distinguished record of supporting research in the polar regions. When the Office of Polar Programs (OPP) was reorganized in 1995, two distinct sections were created—the Antarctic Sciences Section to support research in the Antarctic and the Arctic Sciences Section to focus on science in the Arctic. Within the purview of the Arctic Sciences Section, three programs were established—Arctic System Science, Arctic Social Sciences, and Arctic Natural Sciences (ANS). The ANS program, the focus of this study, is a multidisciplinary effort intended to support work in the space sciences, atmospheric sciences, biology, geology, glaciology, oceanography, and other natural science disciplines.

The scope of the ANS program presented special management challenges, and as a result OPP asked the National Research Council (NRC) for advice to guide the program's future evolution. This report was prepared by a committee carefully constituted to address this task. The report reviews the ANS program —its goals, structure, management strategy, and operation to date. It then considers the defining characteristics of proposals that are suitable for ANS funding and how the program might be best structured to support the broad multidisciplinary nature of its mission. It also contains comments on two other elements of the program's responsibility: the importance of international links and the vital issue of providing logistics support for research in the Arctic. The report concludes with a series of recommendations based on the deliberations of the Committee on Science Priorities for NSF's Arctic Natural Sciences Program and input from the research community.

On behalf of the committee, I would like to thank the staff at the National

Science Foundation for their willingness to respond to our questions and requests for data, especially the staff of the ANS program. We also would like to commend Dr. Odile de la Beaujardiere for her hard work as the program's first manager. Finally, I would also like to thank the committee's members for giving their time and energy to this activity.

John Andrews
*Chair*

# Acknowledgments

Although many people had a role in the preparation of this report and deserve thanks, special appreciation goes to the reviewers. These individuals were chosen for their diverse perspectives and technical expertise, in accordance with procedures approved by the NRC's Report Review Committee. This independent review provided candid and critical comments that assisted the authors and the NRC in making the published report as sound as possible and ensured that the report meets institutional standards for objectivity, evidence, and responsiveness to the study charge. The content of the review comments and draft manuscript remain confidential to protect the integrity of the deliberative process. We wish to thank the following individuals for their participation in the review of this report:

Brian Barnes, University of Alaska at Fairbanks
David Chapman, Woods Hole Oceanographic Institution
David Clark, University of Wisconsin, Madison
Arthur Lachenbruch, U.S. Geological Survey
Debra Meese, U.S. Army Cold Regions Research and Engineering Laboratory
Richard A. Meserve, Covington & Burling, Washington, DC
Ed Myers, National Oceanic and Atmospheric Administration
Walter Oechel, San Diego State University
Glenn Shaw, University of Alaska at Fairbanks

While the individuals listed above have provided many constructive comments and suggestions, responsibility for the final content of this report rests solely with the authoring committee and the NRC.

# Contents

# Summary

The National Science Foundation's (NSF) Office of Polar Programs (OPP) is charged with enhancing our understanding of the Arctic and Antarctic—their harsh environments, systems, resources, and roles in global processes. Within OPP, the Arctic Section is responsible for a variety of programs that seek to better understand the biological, geological, chemical, and sociocultural processes operating in the Arctic, as well as the interactions of ocean, land, atmosphere, biological, and human systems. One of these programs is the multidisciplinary Arctic Natural Sciences program (ANS), which was established in 1995 and began distributing research funds in 1996. With annual funding of about $10 million, ANS provides grants to support state-of-the-art research in the atmospheric sciences, biological sciences, earth sciences, glaciology, and oceanography. The program thus considers proposals in an exceptionally wide range of fields, and while this is scientifically stimulating it is challenging from a management point of view.

Because the program is young and still evolving, OPP asked the National Research Council to examine the ANS program's management and research strategy and provide guidance on how to set research priorities given the diverse scientific issues that fall within its purview. This report is the response of a committee appointed to accomplish that task.

The committee is unanimous in stressing that there is a significant need for a research program with the diverse focus of the Arctic Natural Sciences program, and—despite some inherent overlap with other NSF programs—the committee believes that the ANS program should be NSF's focal point for research in the natural sciences in the Arctic. Individual as well as joint and cooperative proposals should have equal access to funding.

The ANS program is off to a good start, but this is an opportune time to identify areas for its improvement. This report explores the program's strengths and weaknesses, looking at its current budget, portfolio, and management structure. This summary provides a brief overview of the committee's main findings, but readers should see the full text for details.

## PROGRAM SCOPE AND STRUCTURE

In the committee's view, the program's effectiveness can be increased by more clearly defining the ANS mission statement and scientific boundaries, while keeping in mind the ANS' role within the full context of programs within OPP and NSF as a whole. After studying the ANS program and the OPP context, the committee offers the following as a mission statement for the program:

> The mission of the Arctic Natural Sciences Program is to fund cutting-edge research dealing with any aspect of the Arctic's atmospheric, terrestrial, and marine systems. The program focuses on proposals that contribute to understanding and predicting the unique elements and processes that are part of the arctic environment.

The committee suggests that the ANS's broad scientific program be subdivided and managed in three spheres: atmospheric systems, terrestrial systems, and marine systems. These spheres should be seen as administrative aids; their boundaries should be flexible with regard to both subject matter and allocation of dollars. There will be instances of overlap and interaction among the spheres, and times when specific proposals will fail to fit neatly into a category. In these cases, staff should let good judgment rule and assign an appropriate, if not perfect, home. The boundaries are flexible to accommodate varied circumstances and should not be allowed to inadvertently foster competitiveness among the groups; the effectiveness of the ANS program as a whole should always be the key objective.

The first step in determining whether a proposal is suitable for ANS funding is to judge whether the idea falls within program boundaries. The committee proposes three levels of guidelines to help program staff select appropriate proposals. First, the research should focus on one of the following:

- The research addresses natural phenomena, problems, or processes associated with arctic latitudes.
- The arctic environment serves as an irreplaceable natural laboratory or unique source of data or materials for the successful execution of the research.
- The research deals with glaciology—including ice sheets, glaciers, snow, and permafrost—anywhere in the Northern Hemisphere.

Second, the program should be open to:

- research proposals with an arctic focus that are not part of existing targeted, focused programs elsewhere within NSF;
- research proposals considered of higher risk or uncertainty, or perhaps unusually creative or speculative in nature;
- research proposals that require extensive logistical support in the Arctic; and
- research proposals that gather data of use to the international arctic science community and that involve collaboration with international partners.

Third, as a final level of guidance, the committee believes that ANS proposals, in general, can include:

- research that examines processes along latitudinal gradients that include the Arctic;
- research ideas from all scientists, whether or not they have had previous experience in the Arctic;
- research that requires major synthesis of data and theories within the Arctic, or
- research that investigates bipolar processes.

## MANAGEMENT STRATEGY AND RESEARCH PRIORITIES

The fundamental strength of the ANS program is that it is a general program that covers a broad range of topics. It is not theme-focused, nor does this committee believe it should be. The program's flexibility makes it a good opportunity for ideas that are innovative and perhaps even risky. It can be a starting point for scientists with limited experience in the Arctic. Although priority setting will always be a part of decisionmaking where budgets are limited, the selection of specific themes that dictate program direction and are solicited as such should not be the driver behind this program. We do not mean to imply that other NSF programs are not open to newcomers or do not support innovative research; rather, our point is that the ANS programs' breadth can be seen as a strength and used to facilitate such activities.

Because it is hard for program managers to make judgments in fields outside their own, decisionmaking in such a diverse program can be difficult. This problem can be somewhat alleviated by careful selection of staff—people with expertise and experience in a variety of the relevant disciplines; it will be further mitigated over time as staff knowledge increases. In addition, aspects of the decisionmaking process can be designed to enlarge the view brought by the program's managers. For example, setting priorities should involve input from at least three groups: *NSF management*, to be sure that the priorities selected support broader agency goals and strategic planning; the *research communities* that will be requesting funds, to be sure that the priorities selected represent what they believe are the most important and cutting-edge issues; and *representatives of related research programs and agencies*, to be sure there is coordination of effort.

This input can be gained through a one-time exercise like a workshop or special committee, but a more long-term approach would be to use existing activities to solicit ongoing input—that is, use existing mail review and panel review processes to judge proposal quality and gain insight into the level of importance of the work. Other mechanisms to help in priority setting could include Committees of Visitors tailored to the task, subcommittees of the Office Advisory Committee, and town meetings. These processes, combined with input from NSF staff and from other agency staff, should be adequate to reflect the full range of views. It is then the job of ANS staff to synthesize the information and make final judgments about priorities. In general, the setting of priorities for the ANS program should not be a one-time event but a flexible and continuing process.

## INTERNATIONAL AND AGENCY COOPERATION

Since many nations are actively involved in research in the Arctic, international cooperation is vital in optimizing opportunities and cost effectiveness; thus, international collaboration should be encouraged. Although a research proposal cannot be judged solely on whether it has an international dimension, appropriate collaborations across national boundaries should almost always be considered an advantage. Similarly, interagency cooperation is critical to the efficient use of limited resources. Some formal mechanisms exist to encourage communication and cooperation among agencies (e.g., the Interagency Arctic Research Policy Committee) but ANS staff should be encouraged to continue and increase informal interactions with staff in other agencies as well.

## LOGISTICS

Any program charged to support research in the Arctic will inevitably have to deal with the issue of providing logistics support, and OPP needs to address how best to provide such support in the arctic context. This is not to say that arctic logistics support must be equal to the antarctic logistics support provided by the Polar Support Section, because the settings are very different — and, in fact, many in the arctic research community appreciate the flexibility they have to arrange their own logistics. But there is community dissatisfaction with the current approach, where each program manager is expected to juggle some logistics issues.

## PROGRAM DATA AVAILABILITY

As noted, the ANS program is relatively young and so has a limited record of performance. Still, in the course of this study it became apparent that basic information about the program was not readily available. Good recordkeeping is essential to good program management: accurate, basic data on the size and types of awards and other standard information should be kept in computer databases that are accessible to staff and easily interpreted.

## CONCLUSION

In summary, the Arctic Natural Sciences Program is responsible for selecting and funding research in an exceptionally wide range of fields, and this does pose some special management challenges. Given the diversity of fields covered and the volume of proposals received, initial staffing levels were too small but recent changes may have relieved this situation. Regardless of the number of staff assigned to the program, special care will always be needed to ensure that the program fulfills its broad mission—to be the main NSF program responsible for funding cutting-edge research dealing with the Arctic's atmospheric, terrestrial, and marine systems. While it may periodically be appropriate to highlight certain research areas as being of current high importance, the fundamental strength of the ANS program is its breadth, and it should not evolve into a program that is theme-driven. As a general program, ANS provides important opportunities for individual researchers, new ideas, and disciplinary approaches that do not fit into the focused themes that guide many of the other relevant programs.

# 1

# Introduction

The polar regions provide unique research opportunities in a variety of scientific disciplines, from the geology of the deep ocean bottom to the plasma physics of the high atmosphere. The National Science Foundation's (NSF) Office of Polar Programs (OPP) plays a major role in funding and providing logistical support for research to enhance our understanding of the polar regions. Within OPP, the Arctic Sciences Section was established to gain a better understanding of the Arctic's natural and sociocultural processes and of the interactions of ocean, land, and atmosphere. The Arctic Section gives emphasis to an exceptionally rich arena for research: the Arctic is where environmental change is expected to be greatest—and where a warming of a few degrees can profoundly alter the environment by transforming ice to water. It is an ideal place for the study of integrated global systems and of human dimensions. And it is a region with abundant renewable and nonrenewable resources.

OPP's Arctic Section is divided into three main programs: the Arctic Social Sciences program (ASSP), the Arctic System Science program (ARCSS), and the Arctic Natural Sciences program (ANS). The ASSP is intended to support research in the full range of social sciences, including anthropology, archaeology, economics, geography, linguistics, political science, psychology, sociology, and related subjects. The ARCSS program is intended to focus on research related to selected interdisciplinary themes related to understanding the physical, geological, chemical, biological, and sociocultural processes of the arctic system that interact with the total Earth system and thus contribute to or are influenced by global change. The ARCSS research goal is to advance the scientific basis for predicting environmental change on a seasonal-to-centuries time scale and to formulate possible policy options for responding to the impacts of global change.

The ANS program is a multidisciplinary program that is intended to provide grants to support state-of-the-art research in the atmospheric sciences, biological sciences, earth sciences, glaciology, and oceanography in the Arctic. The program's broad mission means that it must coordinate its work with related programs in other parts of NSF, especially the Geosciences Directorate, and, ideally, take advantage of collaborations with other agencies and sometimes other nations. In its first few years of existence, the ANS program has had an annual budget of approximately $10 million and has provided funding for projects in a diverse array of fields. More details about the nature of the program are included in Chapter 2, which provides an overview of the program's portfolio, and in Appendix A.

## THE COMMITTEE'S CHARGE

The ANS program was established in 1995 and began distributing grants in 1996; as a young program, it is still evolving in scope and management strategy. To seek ways to make the program more effective, OPP requested that the National Research Council (NRC) form a committee to examine the ANS program's management and research strategy and provide guidance on how to set research priorities given the diverse scientific issues that fall within its purview. In particular, the committee was asked to:

- suggest improvements to the program's management strategy, including ways to compare proposals in widely diverse fields;
- suggest how to judge which proposals are best suited for the ANS program versus other, related NSF programs; and
- suggest ways to improve interagency and international collaborations related to the ANS mission.

## STUDY METHODS

To conduct this study, the NRC appointed a volunteer committee of nine experts, each selected to bring appropriate expertise as well as the ability to take a multidisciplinary perspective. The committee's members have experience in many aspects of natural sciences as covered by the ANS program, including the atmospheric sciences, oceanography, glaciology, biology, geology, and geophysics. The full committee met twice, once to gather information about ANS history, function, and current management, and later to deliberate on and refine its final report. Considerable work was accomplished via e-mail, telephone conference calls, and an editorial subgroup meeting.

Efforts were made to seek input from the arctic research community. A survey was designed and posted on the World Wide Web to solicit views on the ANS program and on funding for arctic research in general (see Box 1-1). "Town meetings" were held in December 1997 at the annual meeting of the American Geophysical Union and in February 1998 at a major meeting of ocean scientists (see Boxes 1-2 and 1-3). Staff at NSF—and at OPP in particular—were helpful in providing background information, data, and analysis relating to the numbers and types of proposals received and funded and other issues. Outside review was conducted according to standard NRC procedures, and the reviewers provided useful insights that have been incorporated into the committee's thinking.

## ANS PROGRAM HISTORY

Some sense of history is helpful in understanding the ANS program. What is today known as the Office of Polar Programs (OPP) began as the Division of Polar Programs, housed within the Geosciences Directorate. The unit was created, in large part, to help the National Science Foundation carry out its role of supporting a credible U.S. presence in Antarctica, a task that has both scientific and political dimensions. The proposal evaluation process was administered by separate disciplinary science programs (i.e., aeronomy and astrophysics, biology and medicine, geology and geophysics, glaciology, polar oceans and climate systems) and there was a separate unit to fund and organize logistics.

As the office grew and evolved in the 1970s and 1980s, requests for support for research in the Arctic increased and were addressed within the existing disciplinary structure—that is, proposals related to biology in the Arctic went to the Biology and Medicine Program, glaciology proposals regardless of location were handled by the Glaciology Program, and so forth. Still, the historic emphasis on the Antarctic gave at least the perception of inequity.

Another aspect of context important to understanding the U.S. approach to research in the Arctic was passage of the Arctic Research and Policy Act of 1984 (Public Law 98-373). The Act defines what is to be included when we refer to the Arctic[1] and establishes a number of mechanisms to "provide for a comprehensive national policy dealing with national research needs and objectives in the Arctic." Foremost among these are the Arctic Research Commission (ARC), charged to recommend an integrated national research policy in the Arctic and to guide federal agencies in implementing their research programs, and the Interagency Arctic Research Policy Committee (IARPC), charged to survey arctic research conducted by federal, state, and local agencies, universities, and other public and private institutions to help determine research priorities and work with

[1]The U.S. Arctic Research and Policy Act of 1984 defines the Arctic to be all United States and foreign territory north of the Arctic Circle and all United States territory north and west of the boundary formed by the Porcupine, Yukon, and Kuskokwim rivers; all contiguous seas including the Arctic Ocean and the Beaufort, Bering, and Chukchi Seas; and the Aleutian chain.

the Arctic Research Commission on developing national research policy to guide the agencies. IARPC produces a regular 5-year plan for implementing arctic research policy. ARC is composed of seven members appointed by the President, including individuals from academic institutions, representatives of indigenous residents of the Arctic, and others familiar with the needs of private industry in the Arctic. IARPC is composed of representatives of the National Science Foundation, Department of Commerce, Department of Defense, Department of Energy, Department of the Interior, Department of State, Department of Transportation, Department of Health and Human Services, National Aeronautics and Space Administration, Environmental Protection Agency, and others as deemed appropriate. ARC and IARPC were established to be mechanisms for addressing arctic research policy and priorities and indicate the increasing importance of the issue.

In 1995, OPP decided that to have a true long-term commitment to science in both polar regions, it needed to create a structure that gave explicit attention to arctic science, an idea encouraged by members of Congress. OPP was divided into three sections, an Antarctic Sciences Section, an Arctic Sciences Section, and Polar Research Support Section (although this is a misnomer, as the section serves Antarctic research exclusively). The Arctic Sciences Section was organized into various programs: first the Social Science program (ASSP) and Arctic System Science program (ARCSS) were created, and then the Arctic Natural Science program (ANS) was added shortly thereafter.

This committee was not asked to evaluate the past decision to split OPP into Arctic and Antarctic Sections, nor the subsequent decision to organize the Arctic Sciences Section into three subsections not based on disciplines, as was the tradition in the Antarctic Section. Although management needs and staffing issues were the likely motivation for this later decision, the move toward more multidisciplinary thinking is in keeping with current trends in science.

However, the committee cannot help but note that many of the problems faced by the ANS program arise, at least in part, because of this structure. In particular, there is an inevitable overlap between the ANS program and the ARCSS program. Both are, by the normal usage of the phrase, "natural science." Because ARCSS is the more clearly defined of the two, addressing selected interdisciplinary themes related to understanding the arctic system and predictng environmental change, ANS has sometimes been seen as the place for things that do not belong in ARCSS, rather than having its own clear, inclusive definition. If logic alone were to guide structure, the Arctic Sciences Section might have two subsections, Arctic Social Sciences and Arctic Natural Sciences, and Arctic Natural Sciences might have a subunit called Arctic System Science.

But while such hindsight is interesting, it is not helpful in the current effort to improve the ANS program. Instead, in the following chapters the committee turns its attention to describing the existing ANS program's structure and its status and portfolio of projects. We then proceed to suggest improvements to the existing program design so it becomes better able to set priorities and maintain a comprehensive, balanced portfolio of projects.

## BOX 1-1
## ELECTRONIC QUESTIONNAIRE TO GATHER COMMUNITY VIEWS ON THE ARCTIC NATURAL SCIENCES PROGRAM

To obtain input from the scientific community at large, the committee designed a questionnaire to poll the users and potential users of ANS. This questionnaire was posted on the Polar Research Board's homepage, advertised on the ARCUS listserver, and publicized at Town Meetings (see Boxes 1-2 and 1-3) and by word of mouth. The questionnaire had three sections: the first characterized the background of the respondents; the second asked specifics about the respondents' experience and perception of ANS; and the third asked about future directions for ANS.

Although the number of responses received was low (approximately 30), they do provide useful anecdotal insights about how the scientific community views the ANS program. It is not the aim of this summary to document each individual anonymous response, although examples of some of the more pertinent comments are included in Chapter 4 to illustrate particular views and issues.

Approximately 25 percent of respondents were senior scientists, 50 percent were mid-career scientists, 13 percent considered themselves early career scientists, and 12 percent fell into the "other" category, which included graduate students, postdocs, and others including a few respondents outside of the United States. Respondents' NSF funding over the past 5 years had ranged from 0 to 100 percent.

Many of the issues raised by respondents were similar to the topics discussed by the committee during its meetings. However, this electronic venue provided an opportunity to provide specific and often very detailed outside comments. Not surprisingly, a wide spectrum of responses was received ranging from the very positive to real dissatisfaction. In many instances, helpful suggestions were made.

Many of the active researchers obtained funding from other primary funding sources (e.g., NASA, DOE, NOAA, USDA, DOD, USGS, private foundations) as well as from other NSF programs (e.g., ARCSS, EHS). In general, few had been involved in international collaborations or had received funding from other nations.

The questionnaire polled respondents for their views on ways to improve the ANS program and suggestions for future directions. By far, most of the respondents encouraged the ANS program's broad attempt to support multidisciplinary and interdisciplinary research and believed it should continue unconstrained by traditional disciplines and themes. Most felt that an open approach to arctic science had been practiced, but wondered if cross-polar, tropical glaciers, and future ice core studies risked falling through the cracks of existing funding structures. An increase in cost sharing with programs such as ARCSS and ESH was suggested.

Much concern was voiced over the impossible task faced by a sole program manager in managing such a large and diverse program. The use of a scientific panel and heavy reliance on reviewer's comments was strongly urged. There was general concern that logistics in the arctic could become a problem in the future, with approximately half the respondents' research being constrained by high logistics costs.

That so many parallel and common responses were received by the committee from the Town Meetings and this questionnaire lends a certain degree of confidence that a representative spectrum of the scientific community had been consulted.

**BOX 1-2**
**TOWN MEETING**
**December 12, 1997**
**The AGU 1997 Annual Meeting**
**San Francisco, California**

A Town Meeting to gather input from scientists who work in the Arctic was held December 12, 1997 in San Francisco at the American Geophysical Union Annual Meeting. The meeting was well attended: the participants ranged in number from 50 to 75 through the course of the one and a half hour session. They represented diverse disciplines. Four members of the review committee were also present: Drs. B. Hallet, P. Mayewski, J. Morison and R. Smith. NSF staff was invited to attend the first part of the meeting to provide an introduction to ANS and explain the purpose of the committee. After an introduction by current ANS program manager, Odile de la Beaujardiere, and a few comments from Arctic Section Director, Tom Pyle, the NSF staff was thanked and left; the meeting continued with a candid discussion.

To a large extent, the discussion and the range of concerns paralleled those that the committee itself had during its meetings. Many participants expressed concern with the obvious challenge of having a single manager for a broad multidisciplinary program such as ANS. The virtue of having a program unconstrained by thematic and disciplinary boundaries was expressed several times. The merits and challenges of subdividing the program along broad disciplinary or thematic lines were also discussed, but no consensus emerged about how to best accomplish this. Logistics was a significant concern for several participants.

The Town Meeting did provide new light on some issues. In her introduction, ANS program manager Dr. de la Beaujardiere responded to a question about her involvement with major collaborative research efforts and programs by noting that the program's primary concern is for individual principal investigator research efforts with the belief that organized research groups were well supported by other programs (e.g., ARCSS, PALE, etc.). The notion of ANS having a review panel, which would seem particularly beneficial for such a broad program, was brought up but generated considerable discussion of both pros and cons; once again, no clear consensus emerged or could be expected from this relatively short and informal discussion.

Overall the participants seemed primarily interested in learning more about ANS and about ways to improve it, rather than voicing particular concerns about the program based on personal experience. There was no general discontent, but many of the questions raised lacked substantial answers. At the meeting, participants were given only limited and general information about aspects of ANS. They were not given specific data about numbers of proposals, partitioning of funds and success ratios across disciplines, etc. As a result, their response was based upon broader aspects of the program and personal experience.

The Town Meeting provided an opportunity to involve the scientific community in discussing the organization, management, disciplinary coverage, and other general aspects of ANS. However, the dearth of specific data available to the participants made it difficult for them to comment on the balance of allocations according to disciplines and other specific program issues. At the meeting we discussed the definition of ANS, what it includes and excludes, but did not reach clear conclusions. It is likely that many participants left the meeting with considerable uncertainty about exactly what proposals would fare well at ANS, what were the boundaries between ANS and related programs (e.g. ARCSS, ESH), and what would be the impact of major anticipated logistical expenses (e.g., USCGC Healy) on ANS.

**BOX 1-3**
**TOWN MEETING**
**February 11, 1998**
**The AGU 1998 Ocean Sciences Meeting**
**San Diego, California**

The second Town Meeting for community discussion of the Arctic Natural Sciences program was held February 11, 1998, at the American Geophysical Union 1998 Ocean Sciences Meeting in San Diego, California. The committee felt that hosting another session at this meeting was a good idea because many ocean scientists would be attending it instead of the Fall AGU meeting. The Town Meeting was run by committee members Drs. J. Morison and B. Hallet and attended by 17 scientists; because of the small size of the group the discussion format was informal. ANS program manager Dr. Odile de la Beaujardiere again gave a short presentation on the program. She discussed the funding levels and general philosophy of the program, pointing out that it is driven by individual investigator research in any area of natural science pertaining to the Arctic. She indicated that ANS does consider global change oriented proposals if they are not suitable for ARCSS. After her presentation NSF personnel left and community discussion continued.

The group discussion was wide ranging but several themes were emphasized. Most people felt it was good that ANS provides a resource for the single principal investigator proposal. They felt this was a good contrast to ARCSS, which often requires a concerted group effort and dealing with numerous planning committees. ANS provides a unique opportunity for the individuals who are not part of a large program.

In other discussions, the logistics of working in the Arctic proved to be a major concern, not only the availability of platforms but the manner in which arctic logistics is funded. The universal concern was that logistics costs for arctic work had to be borne by the science budgets, and especially in the case of individual principal investigator proposals actually have to appear in the principal investigator's budget. This it was felt put Arctic researchers at a disadvantage compared to those in other programs where some dedicated support was available.

Most people thought ANS and the Arctic Sciences Section in general were understaffed. Several people complained that it was getting hard to reach their funding officers because they were so busy. While there was consensus on the problems, there was not a consensus as to answers. Some mention was made of using the logistics support of thematic programs such as ARCSS to help individual researchers. Some discussion was given to a discipline-oriented breakdown for ANS. There was not universal enthusiasm for this idea, however, because it could result in the program being divided into too many small pieces to put together a good program in any one area.

# 2

# Budget and Portfolio

To provide guidance about future directions for the Arctic Natural Sciences program (ANS), the committee examined the budget and portfolio of the current program and considered its history to gain perspective on scope, strengths, and weaknesses.

According to the most current data available when this report was being written, the ANS program has an overall budget of just over $10 million (Table 2-1). According to Office of Polar Programs estimates, the Arctic Sciences Section and Antarctic Sciences Section receive about the same base funding: just over $30.5 million. A key difference is that the Arctic Section's $30.5 million includes all logistics costs, whereas antarctic science is supported by a separate logistics section, the Polar Research Support Section, which has a total budget of about $162.4 million. (It is difficult to determine how much of this directly supports research compared to the other goals related to the U.S. presence in the Antarctic.) The difference in logistics support and the many debates surrounding equity of support at the two poles is beyond the scope of this study, but the committee heard repeated concerns about this issue during our outreach activities —an indication that some serious attention is merited.

The ANS budget of approximately $10 million seems adequate in light of the number of proposals being received (the "proposal pressure"), and the funds appear to have been allocated effectively during these first start-up years. The ANS base budget is enhanced by contributions from other National Science Foundation programs, where there is overlap of interests and a decision is made to seek joint funding. Contributions from other programs to ANS amounted to about $3.1 million in Fiscal Year (FY) 1997 (Figure 2-1). Of this, 62 percent was

TABLE 2-1 OPP Operating Accounts, FY 1997 Plan

| | |
|---|---|
| OPP Front Office | $851,000 |
| Arctic Sciences Section | |
| Arctic Research & Policy Support | 263,764 |
| General Arctic Research Support | 414,965 |
| Arctic Logistics | 2,831,272 |
| Arctic System Science | 13,564,292 |
| Arctic Social Sciences | 1,406,953 |
| Arctic Information & Advisory | 100,000 |
| Arctic Research Commission | 500,000 |
| **Arctic Natural Sciences** | **10,316,154** |
| Academic Research Infrastructure | 1,200,000 |
| | 30,597,400 |
| Polar Research Support Section | |
| Science Support | 4,607,497 |
| Operations & Science Support | 94,622,387 |
| Logistical Support | 62,600,000 |
| Environmental Coordination Activities | 565,000 |
| | 162,394,884 |
| Antarctic Sciences Section | |
| Antarctic Environmental Research | 459,820 |
| Antarctic Biology & Medicine | 7,174,523 |
| Antarctic Geology & Geophysics | 5,269,689 |
| Antarctic Oceans & Climate Systems | 4,420,000 |
| Antarctic Aeronomy & Astrophysics | 3,064,073 |
| Antarctic Glaciology | 4,247,233 |
| Antarctic Coordination & Information | 590,998 |
| Science & Technology Center | 3,929,093 |
| Academic Research Infrastructure | 1,200,000 |
| | 30,355,429 |
| Total | $224,226,790 |

SOURCE: NSF, 1997a.

contributed from "initiatives," cross-directorate programs such as Life in Extreme Environments, Major Research Instrumentation (MRI), and the Plasma Physics joint initiative with the Department of Energy. Some of these are special one-year initiatives, such as the Plasma Physics initiative; others, such as MRI, last for many years. Other areas of NSF also contribute funds to support ANS projects when their interests overlap, including 18 percent from other programs within OPP, 11 percent from the Atmospheric Sciences Division, 6 percent from the Engineering Directorate, and 3 percent from the Ocean Sciences Division. In reverse, the dollars given by ANS to other NSF programs was $248,000; this was

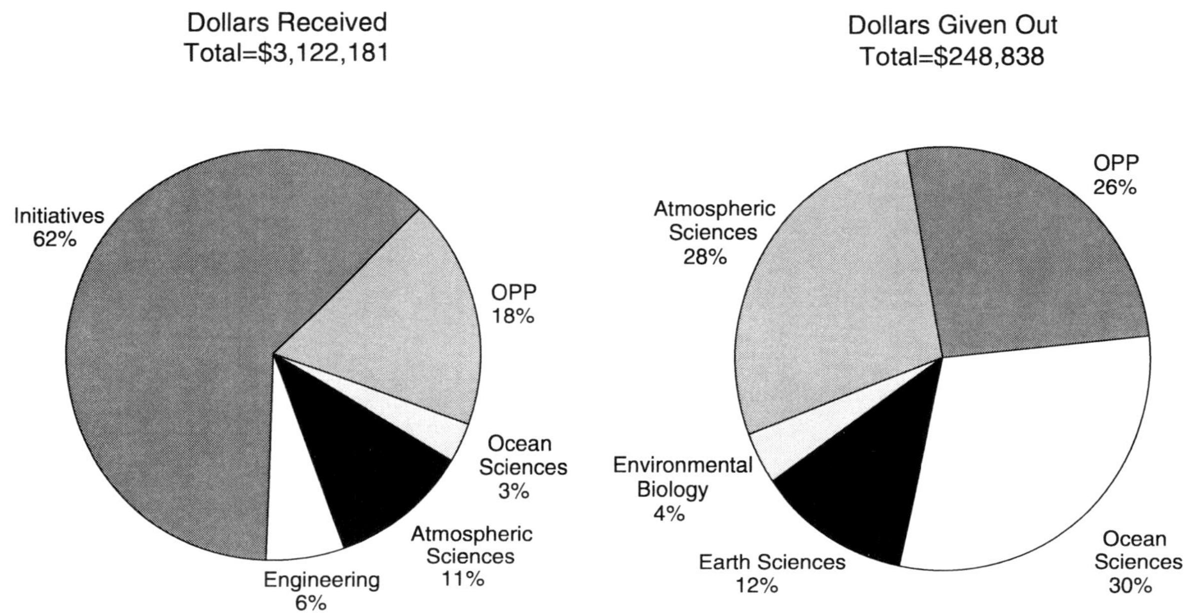


FIGURE 2-1 Dollars received from other NSF programs and dollars provided to other NSF programs, FY 1997.

about evenly divided among OPP, Ocean Sciences, and Atmospheric Sciences, with smaller contributions to the Earth Sciences and the Environmental Biology Divisions.

Since its first year of awards in FY 1996, the ANS program has received 227 proposals: 87 in FY 1996 and 140 in FY 1997.[1] Thirty-four proposals were funded in FY 1996, a success rate of 39 percent. In FY 1997, 45 proposals were funded, a success rate of 32 percent. As best as can be estimated, before the Arctic Section was created the arctic portion of the combined antarctic/arctic budget devoted to areas now covered by the ANS program was about $8.6 million. In 1997, the second year of the ANS program, its budget grew to $10.3 million.

A few words of caution are needed regarding the data presented in this chapter and elsewhere in the report. First, the ANS program does not actually allocate funds by discipline, and all the breakdowns shown are approximate. The committee requested information from ANS staff to help us understand the nature and scope of the program, and we asked that this be provided by discipline so we could compare it to data before 1996 when arctic proposals were handled by the Antarctic's disciplinary programs. The data used, however, are not perfectly parallel. In fact, they show two different characterizations—one matching the old disciplinary programs pre-1996 and a variation provided by ANS staff that includes biological oceanography as a separate category because that is a significant component of ANS activities. The committee had no way to reconcile these different categorizations nor to confirm that proposals were counted in the appropriate disciplinary category.

Second, because the program has existed for only two years, the available data sets are quite small, and the specific numbers cited should not be given too much credence. Also, it proved to be fairly difficult for staff to develop summary data, indicating that recordkeeping and analysis have been limited during the program start-up period. Similarly, changes indicated by program data over the two years cannot be interpreted as significant; where we point out changes, we do so in hopes of identifying potential issues to be watched in the future.

Given these caveats, Table 2-2 summarizes the program in FY 1994 and FY

[1] All program data were provided by OPP staff and are current as of March 1998; to limit our analysis to materials we had an opportunity to discuss face-to-face, the committee did not consider new materials after this date. Because 1996 was a transition year, our analysis focuses on 1995 as representative of "pre-ANS" data and generally treats 1997 as the first full year of ANS data. We have not attempted to update data if new information became available after the committee's final meeting. For informational purposes, note that 130 proposals were received for FY 1998 consideration, with 23 expected to be funded; this is a success rate of 18 percent, a sizable drop from earlier years. Because this occurred after the committee's final meeting, we are not able to offer insights about the reasons for the change.

TABLE 2-2 Comparison of Funding by Disciplines, Before and After Establishment of ANS

| | Pre-ANS | | | | Post-ANS | | | |
|---|---|---|---|---|---|---|---|---|
| | 1994 | | 1995 | | 1997 | | 1997-New Awards | |
| Discipline | $1000s | % | $1000s | % | $1000s | % | $1000s | % |
| Biology/medicine | 2,450 | 30 | 2,450 | 28 | 3,146 | 31 | 2,311 | 47 |
| Oceans/meteorology | 1,563 | 18 | 1,570 | 18 | 1,895 | 19 | 624 | 13 |
| Glaciology | 2,150 | 25 | 2,257 | 26 | 1,491 | 15 | 672 | 14 |
| Geology/geophysics | 1,270 | 16 | 1,345 | 16 | 1,870 | 19 | 410 | 8 |
| Aeronomy/astrophysics | 1,050 | 13 | 1,050 | 12 | 1,487 | 15 | 812 | 16 |
| Miscellaneous | 0 | 0 | 0 | 0 | 127 | 1 | 128 | 1 |
| TOTAL | 8,480 | | 8,672 | | 10,019 | | 4,957 | |

NOTE: Percentages represent proportion of total funding dollars. Some variability in funding is to be expected each year. The committee does not wish to imply that the percentages allocated to each discipline before the creation of ANS are somehow target or desired funding levels; they are, however, baseline data useful in understanding the scope of the program.

SOURCE: NSF, 1997a.

1995 (when arctic research was covered by the antarctic disciplinary programs) and in FY 1997, showing the amount spent and percentage of total funds allocated by discipline. The table indicates that, in general, the proportion of funds allocated by discipline has remained fairly constant despite the changeover in program organization. The greatest relative change in funding was in glaciology, which declined from 26 percent of the budget in FY 1995 to 15 percent in FY 1997. Biology increased its share of the budget, rising from 28 to 31 percent over the period. Funding shares for proposals in geology and geophysics and in aeronomy and astrophysics also rose (from 16 to 19 percent and from 12 to 15 percent, respectively).

Table 2-3 shows the average award size, number of awards in 1997, and percentage of total awards made per discipline. Note that a single large award to an important multidimensional ecology initiative (more than $1 million) skews the data in favor of biology. If this award is taken out of the calculation, the average grant size becomes about $130,000. There is considerable variability in the average awards by discipline, ranging from about $64,000 per grant (miscellaneous) to as high as $194,000 (biology/medicine). Two disciplines receive about $123,000 and two receive just over $80,000 per grant. Although these variations are interesting, the shortness of the program's history makes it impossible to know if they are significant.

Table 2-4 shows the number of proposals received by discipline, the percent this is of the total received (140), the number and percent actually awarded, the percent of successful proposals per discipline, the percent of total funds allocated by discipline, and a ratio calculated to indicate the percentage of proposals received in each discipline to the percent awarded in that discipline. A ratio of close to 1 indicates general agreement between percentage submitted and percentage accepted. For example, glaciology proposals made up 11 percent of the total proposals received by ANS in FY1997 and was the subject of 12 percent of the awards made, for a ratio of 1.09; this example shows a close relationship between

TABLE 2-3 Details of Awards Made, FY 1997

| Discipline | Average Award Size ($) | Number of Awards | Percent of Total Awards by Discipline (%) |
|---|---|---|---|
| Biology/medicine | 194,277 | 12 | 27 |
| Oceans/meteorology | 82,222 | 12 | 27 |
| Glaciology | 123,491 | 6 | 13 |
| Geology/geophysics | 85,109 | 5 | 10 |
| Aeronomy/astrophysics | 123,962 | 8 | 18 |
| Miscellaneous | 63,881 | 2 | 5 |
| Overall program | 124,532 | 45 | 100 |

SOURCE: NSF, 1997a.

TABLE 2-4 Proposals Received Compared to Proposals Awarded by Discipline, FY 1997

| Discipline | Proposals Received | | Awards Made | | Ratio of % Awards Made to % Proposed[a] | Percent Successful Proposals | Percent Total Funds |
|---|---|---|---|---|---|---|---|
| | No. | % | No. | % | | | |
| Aeronomy/Astrophysics | 21 | 15 | 6 | 14 | 0.93 | 29 | 16 |
| Geology/Geophysics | 20 | 14 | 5 | 10 | 0.71 | 25 | 8 |
| Glaciology | 15 | 11 | 6 | 12 | 1.09 | 40 | 14 |
| Oceans/Meteorology | 35 | 25 | 10 | 23 | 0.92 | 29 | 8 |
| Biological Oceanography | 23 | 16 | 11 | 25 | 1.56 | 48 | 26 |
| Biology/Medicine | 24 | 17 | 6 | 12 | 0.71 | 25 | 25 |
| Miscellaneous | 2 | >2 | 2 | 4 | 2.00 | 100 | 3 |
| TOTALS | 140 | 100 | 44 | 100 | | 31 | 100 |

SOURCE: NSF, 1997a.

[a]Table 2-4 shows the number of proposals received by discipline, the percent this is of the total received (140), the number and percent actually awarded, the percent of successful proposals per discipline, the percent of total funds allocated by discipline, and a ratio calculated to indicate the percentage of proposals received in each discipline to the percent awarded in that discipline. A ratio of close to 1 indicates general agreement between percentage submitted and percentage accepted.

proposals received in a discipline and awards made by the program. Biological oceanography proposals, on the other hand, made up 16 percent of the total proposals received but were the subject of 25 percent of the awards made, for a somewhat high ratio of 1.56. Another way to look at this is by success rate: biological oceanography proposals had a success rate of 48 percent, which is higher than the average ANS success rate of 31 percent. The relatively high rate of success may be associated with the high quality of proposals submitted under the auspices of the North Open Water (NOW) project, which has a rigorous procedure for developing proposals aimed at the biological importance of polynyas in the Arctic, or other such coordinated efforts which the committee cannot determine from the information available.

Table 2-4 also shows Biology and Geology/Geophysics are slightly skewed (both have ratios of percent awards made to percent proposed of 0.71). The individual numbers in this analysis may not be meaningful, because of our limited data, but they would be if this were a trend that continued. The percent of funding for the new projects is in the final column of Table 2-4. Almost 50 percent of the total new dollars is going to biology. Again, this is skewed by one large project, which is acceptable within the ANS program and in fact is likely to occur periodically given the ANS program's mission. But even accounting for the decision to fund one large project in any field, care must be taken that over time such decisions do not diminish opportunities in other disciplines. The goal is not absolute equality in numbers of proposals or money awarded, but equity or fairness in a broader sense, perhaps as typified in the allocations originally "given" to ANS by the old disciplinary entities (Table 2-2) and as influenced by proposal pressures, priorities determined through specific strategic thinking, leveraging funds important to international activities, and careful decisions to fund occasional large, complex proposals that clearly fit the ANS mission.

Regarding the percent of successful proposals, as shown in Table 2-4, the committee notes that proposal pressure alone is no indication of quality, and low acceptance rates should not be misinterpreted as a sign of quality. Rather, extremely low acceptance rates indicate that something is amiss in the program and leads to the wasting of researchers' time as they prepare proposals that ultimately have little likelihood of success. An acceptance rate in the 30 percent range, as was typical of the ANS program in its first two years, seems adequate. Significant drops might be indicators of potential problems or imbalances between new and carry-over projects and should spark oversight from the Arctic Section Head. Regarding quality, it is too soon in the program's history to determine number of peer-reviewed publications that have resulted from the grants or otherwise try to evaluate the impact of the grants, but this issue should be considered with care as the program matures and builds a record of completed projects.

# 3

# Management Strategy and Structure

As noted earlier, the organization of the Arctic Sciences Section was designed in response to a variety of management pressures and as a counterpart to the Antarctic Sciences Section. Both manage approximately the same amount of research dollars—about $30.5 million—although the arctic budget supports both research and research logistics, while antarctic logistics are covered separately. Figure 3-1 is an organizational chart for the Office of Polar Programs, current as of February 1998. Several differences in staffing and structure between the two science sections are apparent. For example, the Antarctic Section has 16 positions, including one section head and seven science program managers plus a number of other staff directly involved in managing the research portfolio. (The Antarctic Sciences Section is also supported by 17 staff members and various contractors in the Polar Research Support Section, which provides a variety of logistics services.) Although the position terminology between the two sections is not parallel, the Arctic Sciences Section had 12 positions, including one section head and seven people involved in managing the research portfolio. (This includes one person added to the ANS staff in late 1997 after the committee began its work.)

For science management in the Arctic Sciences Section, there are four temporary (or "rotating") positions, including the section head and the Arctic Natural Sciences (ANS) program manager, and only two permanent positions, one of which was added to support the ANS program during the course of this assessment. In comparison, the Antarctic Sciences Section has six permanent science program managers and a permanent section head. Given the similarity of their

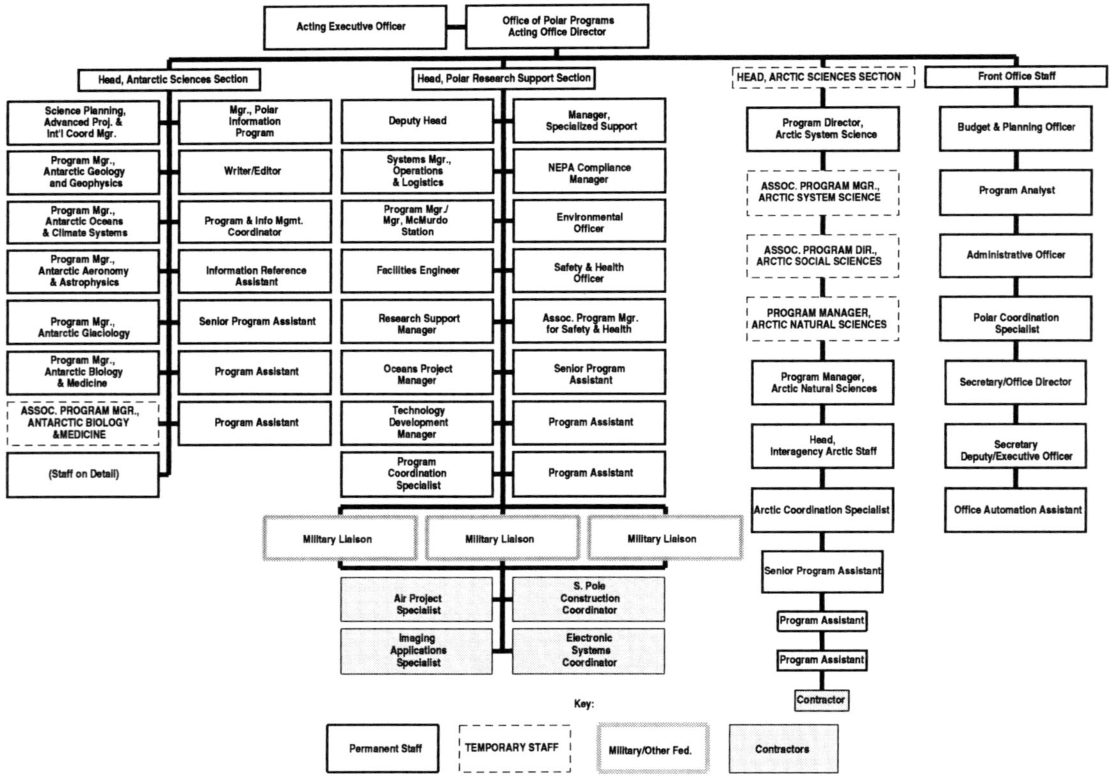


FIGURE 3-1 Office of Polar Programs organizational chart (current as of February 1998).

science budgets, the Arctic Section is understaffed and, moreover, is characterized by a high number of temporary employees.

Although there are benefits to rotating employees, including the influx of new ideas, the high number of temporary staff in the Arctic Sciences Section has implications for continuity of programs and implementation of strategic thinking. Short-term employees are likely to be less effective at certain tasks because it takes time to get up to speed in such a diverse program (in terms of scientific issues and the communities of researchers involved); furthermore, they end up leaving just as they become knowledgeable and effective. Because of the rigid timing of the proposal request cycle, short-term staff will be less able to develop and implement a vision of where the program should head. Short-term employees also may have more difficulty knowing and coordinating with the many other people within the National Science Foundation (NSF) and in other federal agencies with arctic research programs, and such coordination should be an important part of the job for ANS staff. Finally, lack of permanent staff in key program manager positions could be perceived as a lack of long-term commitment to science in the Arctic at the Office of Polar Programs.

## EVALUATION OF CURRENT STRUCTURE AND SCOPE

### Strengths

Although this committee has many suggestions for improving the effectiveness of how OPP manages the arctic side of its portfolio, we should to note that overall the current OPP Arctic Sciences Section has done a credible job of supporting arctic science within the structure imposed upon it during its formation. Although we must inevitably find ways to segment and define scientific categories, or its management becomes overwhelming, science in general is moving away from the neat disciplinary boundaries of the past. Thus the move away from this in the Arctic Sciences Section is not unreasonable.

As for the ANS program, it has shown—in its short history—some distinct strengths:

- The Arctic Natural Sciences program provides a home for a broad range of potential research issues dealing with the atmosphere, biology, geology, and oceans in the Arctic. The program's breadth and flexibility allows it great range in selecting important research.
- Arctic research benefits from these topics being collected together into a single program, because this provides a potential mechanism for scientists from different disciplines to act collectively.
- The ANS program is funded at an adequate level to carry out its mandate, and funds during the first years of operation seem to have been distributed appropriately.

- The ANS program is well suited to fund individual and small team initiatives because it is not tied strongly to any thematic program. Such initiatives are increasingly difficult to fund elsewhere.
- Given its solid funding and flexibility, the ANS program offers a good opportunity to accommodate new—and sometimes even large—ideas. Some caution is necessary in the future, however, as new and perhaps costly opportunities like the new ice-capable research ship (U.S. Coast Guard Cutter Healy) come into service.

### Weaknesses

This committee was charged with evaluating the ANS program so we can provide guidance about future directions, and thus our focus inevitably falls on problems and areas where improvement is possible. The committee notes that the ANS program, and in particular its relationship to other programs within OPP and NSF as a whole, is in some ways still an "experiment in progress" that can benefit from candid feedback. In this vein, we offer the following observations:

- Although the antarctic/arctic split seems irreversible, there is no science-based reason for separating the poles in administration of polar research. What was lost in the process is some focused disciplinary support. More importantly, management and dedicated funding for the logistics of polar science were divided up, to the detriment of the arctic side.
- There is some overlap between the ANS and Arctic System Science (ARCSS) programs. At times, the conditions for inclusion in ANS seem to be "if it's not ARCSS, it's ANS," but this can lead to some confusion.
- The current size of the ANS program and broad diversity of subject matter covered are too great for the staff allocated (although some changes to ameliorate this problem have occurred as this report was being written). The lack of permanent staff in key positions and lack of parity of permanent positions between the Antarctic Section (where most positions are permanent) and Arctic Section (where few positions are permanent) are problems that need to be addressed, albeit within the broader NSF context because of overall limitations on staff levels.
- It is and will continue to be difficult to set priorities for funding when subject matter is diverse. Similarly, it is difficult to use a normal review panel to help in ANS-wide decisionmaking, unless multiple panels are used or special effort is made to design a multidisciplinary panel. Review panels were not universally liked by the town meeting participants, so some thought should be given to how they are used if this route is followed.
- Given the great breadth of subject matter considered in the ANS program, it is and will always be difficult to find staff who can deal fairly and competently with the full range of proposals received.
- Arctic research can be constrained by the lack of logistical support. New

ships are likely to be a heavy burden on the program in the future; this is discussed further in Chapter 5.

- The relationship of the ANS program to other programs that also fund work in the Arctic is not entirely clear to prospective researchers. At one point, it was proposed that there be a group known as "Arctic Affiliates," that is, a group of NSF staff with arctic interests that would meet periodically to provide advice and help review proposals. It would include key staff from the Arctic Section and from units outside the OPP, such as the Earth System History program (ESH), Atmospheric Sciences Division (ATM), and Earth Sciences Division (EAR). The Arctic Affiliates concept has not been taken full advantage of, although an informal approach has developed where the program manager seeks specific input from specific people as necessary. This informal approach seems most useful when the underlying personal relationships are good, and less useful without that foundation. If such a group was operating, consideration might be given to adding members from other agencies funding research in the Arctic, such as the National Oceanic and Atmospheric Administration (NOAA), Office of Naval Research (ONR), and National Aeronautics and Space Administration (NASA), to assist in coordination with these agencies and benefit from their links with international activities.

## EVALUATION BY 1997 COMMITTEE OF VISITORS

Periodic assessments of program success and problems are extremely important for ensuring sound program management, and the Arctic Sciences Section is to be commended for taking steps to evaluate and modify the ANS program while it is still evolving. Other mechanisms, too, can provide periodic input and guidance. One such mechanism, although not targeted exclusively on the ANS program, was a 1997 Committee of Visitors that was charged with evaluating the OPP science program (NSF, 1997b). (Also see Box 3-1.) The report of that Committee of Visitors provided an overall assessment of OPP and looked specifically, albeit briefly, at each OPP program. The following comments from the Committee seem particularly relevant:

- "The Committee [of Visitors] found the science programs within OPP to be in excellent shape . . . some issues require attention, and areas of improvement have been identified. But the strong impression that the Committee formed after two and a half days of examining material is one of a professional, efficient, and hard-working group."
- "OPP is unique within the [National Science] Foundation in that it is not organized along scientific disciplines, but rather along geographic areas. This presents both challenges and opportunities. The challenges include, for example, a dauntingly broad mandate, from astrophysics to microbiology, from ancient climate change to social science. Constant communication among Program Man-

## BOX 3-1
## LESSONS FROM PAST REPORT REMAIN RELEVANT

In 1977, the National Research Council published a report titled *An Evaluation of Arctic Programs Supported by the National Science Foundation.* Two decades later, some of the comments and findings of that report remain current. The report notes, for instance, that "Research in the Arctic embraces most of the basic disciplines and ranges in scope from projects by individual investigators to complex multiagency interdisciplinary programs of several years' duration" and that "a strong U.S. arctic basic research capability requires adequate support for meritorious research in each of the broad areas: earth sciences, ocean sciences, atmospheric sciences, biological sciences, and human sciences." With the exception of the human sciences, which have a separate program under the Office of Polar Program's current structure, all these elements are now part of the Arctic Natural Sciences program. In addition, the report emphasizes the importance of providing funding for both large programs and smaller individual grants, again an important element of the current ANS program.

The report observes that there are substantial fundamental research questions that are best or uniquely pursued in the Arctic—questions about understanding long-term climate changes, the dynamics of the Arctic Ocean, and energy inputs from the magnetosphere; the biota and geochemistry of the Arctic Ocean and its surrounding terrestrial environments; permafrost, arctic hydrology, ice sheets, and ecosystem processes; and the nature of adaptations of species to the harsh environment. These questions are still relevant today. The report concludes that proposals for basic research in the Arctic are better addressed through specific arctic programs and not in competition with all other proposals in the regular disciplinary research programs at NSF because of the special logistics needs and associated higher costs, the increased human risk involved, and the relatively small community of specialized scientists involved.

The 1977 report clearly establishes the importance of promoting "international cooperation through science" and that many arctic processes and issues of interest have global significance and can be more productive and cost-effective with international cooperation and collaboration. The report recommends—much as this report does—that NSF vigorously pursue opportunities to work with other nations in the Arctic and expand its support for joint research activities. It also notes that close interaction between all the federal agencies interested in arctic research is mandatory "to achieve the maximum contributions to science [and] to provide answers to problems of the mission agencies" and it encourages NSF to use its leadership role in the Interagency Arctic Research and Policy Committee to improve coordination among agencies.

SOURCE: NRC, 1977.

agers, both within and outside of OPP, is essential to managing such breadth. The opportunities, however, are equally impressive. The Office's experiment with the interdisciplinary organization of the Arctic Section, while some administrative problems remain to be solved, is speeding the process of integrating physical and social sciences in the Arctic, a critical partnership needed to address society's interaction with, and adaptation to, global change."

- "The use of temporary employees in the interdisciplinary arctic programs seems dangerous, as these Programs require the broader and longer term vision of permanent Program Managers. This situation seems particularly incongruous given the larger number of permanent Program Managers in the antarctic programs, disciplines which lend themselves more readily to rotating positions. In particular, the use of one rotating person to manage all of ANS seems unwise. The Committee could not help but conclude that a fair distribution of grants among the widely varying fields represented in this Program is tremendously difficult, despite the considerable efforts of the Program Manager. Correcting this situation as soon as possible . . . should be a high priority in the Office."

The Committee of Visitors went on to provide comments on each OPP science program, and regarding ANS the group noted that proposals were processed fairly and promptly and that the program was participating in special NSF cross-directorate research initiatives. According to the Committee, the program manager's recommendations were extremely well documented and were appropriate in terms of scope of budget and duration of the project. It noted that the size and number of awards and subject matter distribution and diversity of principal investigators was generally average for NSF, although ANS had an emphasis on smaller projects that seemed appropriate given that ARCSS is tailored to large projects. The Committee found it difficult to assess whether ANS was encouraging high-risk proposals, and noted that at times it was not clear whether a review panel had been consulted; given the interdisciplinary nature of ANS, the Committee noted that use of large panels is commended. Regarding the appropriateness of how resources are allocated in the program, the Committee noted that the ANS program manager was trying hard and doing the "most difficult management job in polar programs." Regarding OPP programs overall, the Committee emphasized the need to take a leadership role in coordinating basic research in the Arctic and the need for better and quicker access to statistical information regarding proposals.

## DOES THE CURRENT ANS STRUCTURE MAKE SENSE?

The issue of whether the current ANS structure makes sense can be evaluated by asking the question: Are there alternative structures that would better serve the community of researchers who now submit proposals to ANS? If there are, are they better structures, in the sense that they allow program management to focus on smaller, more manageable pieces of the overall portfolio or make use of other mechanisms to adequately cover the entire breadth of the program?

One obvious alternative model is to divide ANS into a series of disciplinary units, and thus make it parallel to the Antarctic Sciences Section. Within the Antarctic Section there is considerable sharing and cooperation between the programs so that the boundaries of the programs are fairly transparent, which allows multi- and interdisciplinary research. To institute a disciplinary approach in the Arctic Section, however, would require a number of additional staff and could affect the ARCSS program, which is well-established. As noted earlier, it is a reductionist approach that is less well suited to today's issues than a systems approach.

As currently constituted, the ANS program serves scientific objectives not readily addressed in other ways, and the virtues of any program restructuring must be weighed carefully against the confusion and costs of implementation. This committee, in our discussion of guiding principles for the program in Chapter 4 and our recommendations in Chapter 5, has tried to suggest realistic changes that might bring real benefits without wholesale restructuring of the Arctic Sciences Section or imposing staff or financial demands that are dramatically outside what is currently available.

# 4

# Mission and Guiding Principles

To improve the effectiveness of the Arctic Natural Sciences program (ANS), the mission statement and scientific boundaries need to be better defined, keeping in mind the role of the ANS program within the Office of Polar Programs (OPP) and in relation to National Science Foundation's (NSF) other units. Once its mission and boundaries are more clearly defined, guidelines can help staff members determine the relevance and appropriateness of a given proposal to ANS versus other OPP or NSF programs. Program managers can then use existing mechanisms to ensure that the research selected for funding is both of high quality and high importance.

The current ANS mission is to support "research in glaciology and in the atmospheric, biological, earth, and ocean sciences. . . Areas of special interest include: marine and terrestrial ecosystems, atmospheric chemistry, exploration of the Arctic Ocean, as well as arctic geological and glaciological processes" (NSF, 1998). After studying the ANS program and the OPP context, the committee offers the following as a mission statement for the program:

> The mission of the Arctic Natural Sciences program is to fund cutting-edge research dealing with any aspect of the Arctic's atmospheric, terrestrial, and marine systems. The program focuses on proposals that contribute to understanding and predicting the unique elements and processes that are part of the arctic environment.

**BOX 4-1**
**THOUGHTS ON ANS PROGRAM BOUNDARIES[1]**

"The beauty of the arctic programs generally is their ability to transcend the usual boundaries of discipline or category. . . . I think that ANS shouldn't fall into the trap of dividing things up too formally."

"[The] broad scope is good, provided that the reviewer's comments are respected by the panel and the program personnel in areas in which they have little expertise."

[1]The information in this and the other boxes in this chapter is extracted from responses received to the committee's electronic survey, posted on the Polar Research Board homepage to obtain input from the scientific community. About 30 people responded. These quotes are included to illustrate the views provided; all responses were anonymous.

## GUIDELINES FOR SELECTING PROPOSALS

The first step in determining whether a proposal is suitable for ANS funding is to judge whether the idea falls within the boundaries of the program. Thus the committee set out to develop a set of guidelines that could be applied by ANS staff to help them make these judgments. The committee began by examining the general nature of the proposals considered and funded in the past and by considering the mission statements of ANS and other NSF programs that include elements of arctic natural science. The committee then brainstormed an extensive list of possible criteria, refined and reduced the list, and reached consensus on the most important—or first level—guidelines. These guidelines are not necessarily exclusive to the ANS program, but in total can help managers in their decisionmaking process.

According to the committee, research funded by the ANS program should focus on one or more of the following:

- The research addresses natural phenomena, problems, or processes associated with arctic latitudes.
- The arctic environment serves as an irreplaceable natural laboratory or unique source of data or materials.
- The research deals with glaciology—including ice sheets, glaciers, snow, and permafrost—anywhere in the Northern Hemisphere.[2]

[2]Historically, most glaciology work has been housed within OPP regardless of whether it was conducted in the polar regions, with a few notable exceptions such as funding for the Greenland Ice Sheet Project 2 (GISP2), and the committee's suggestion continues this pattern. Glaciological studies conducted in the southern hemisphere should be considered by the Antarctic Section's Glaciology Program.

Second, as decisionmaking about the appropriate home for a proposal continues, the program should be open to:

- research proposals with an arctic focus that are not part of existing programs elsewhere within NSF;
- research proposals considered of higher risk with respect to successful exploration of the hypotheses outlined, and perhaps unusually creative or speculative in nature;
- research proposals that require extensive logistical support in the Arctic; and
- research proposals that gather data and are of use to the international arctic science community and that involve collaboration with international partners.

Third, as a final level of guidance, the committee believes that ANS proposals can include:

- research that examines processes along latitudinal gradients that include the Arctic;
- research ideas from all scientists, whether or not they have had previous experience in the Arctic;
- research that requires major synthesis of data and theories within the Arctic; or
- research that investigates bipolar processes.

The first level guidelines basically ensure that the proposal addresses the basic mission of the ANS program. The second and third levels expand and elaborate on the types of projects that are most likely to be appropriate to ANS, emphasizing the areas that the committee believes are the program's greatest strengths. For instance, one important element of ANS is that it is available to cover proposals that simply do not fit into other, thematically focused programs, and this, in turn, makes the ANS program especially appealing to young scientists and newcomers to arctic research. Because of the program's breadth, the committee sees it as especially welcoming of high risk, speculative, or innovative proposals, which might be difficult to place in more structured or planned programs. Some of the other guidelines—openness to proposals that require extensive logistics or require major synthesis of data or theories or that investigate bipolar processes—are included because the committee believes the ANS program should be welcoming of such proposals as they might not be easy to place in other, related NSF programs. We do not mean to imply that other NSF programs are not open to newcomers or do not support innovative research; rather, our point is that the ANS program's breadth can be seen as a strength and taken advantage of to facilitate these kinds of activities.

It is recognized, both by OPP management and this committee, that there is potential overlap between the ANS program and the Arctic System Science

(ARCSS) program. As noted in Chapter 1 and presented in detail in Appendix A, the ARCSS program supports interdisciplinary research to understand the physical, geological, chemical, biological, and sociocultural processes of the arctic system that interact with the total Earth system and thus contribute to or are influenced by global change. ARCSS's goal is to advance the scientific basis for predicting environmental change on a seasonal-to-centuries time scale and formulating policy options to respond to the anticipated impacts of global change on humans and societal support systems. ARCSS has four themes: understanding global and regional impacts of the arctic climate system and its variability; determining the role of the Arctic in global biogeochemical cycling; identifying global change impacts on the structure and stability of arctic ecosystems; and establishing the links between environmental change and human activity. ARCSS has a number of clearly defined areas in which proposals are encouraged, including Ocean/Atmosphere/Ice Interactions, Land/Atmosphere/Ice Interactions, Paleoenvironmental Studies, and Human Dimensions of the Arctic System. ARCSS directs much of its support to large, integrated research projects that develop out of a careful planning process, although it will consider proposals from individuals and small groups of investigators.

The interdisciplinary nature of ARCSS is well-suited to its mission, and it is an important program. Given that both the ARCSS program and the ANS program support research in an overlapping suite of disciplines, there will always be some overlap of interests. The clear thematic approach used to guide the ARCSS program is perhaps the most important factor for differentiating the programs: if a subject has been determined to be a priority for the ARCSS program, then the ANS program should continue to steer appropriate proposals in that direction and avoid duplication of efforts. ANS should not be precluded from supporting work related to global change, the underlying thrust of ARCSS, because such research is clearly a part of the ANS mission. It differs from ARCSS insofar as the global change directive of ARCSS is handled in a coordinated, interdisciplinary fashion. ANS should be seen as the broad home for natural science research in the Arctic, but it is essential that program managers understand the full range of arctic research activities that are occurring elsewhere within the NSF structure and coordinate with those activities. Given the three main components of the Arctic Sciences Section—ANS, ARCSS, and Arctic Social Sciences—the committee understands why ANS has had the appearance of "if a proposal does not belong in ARCSS, it must belong in ANS." While this can create some confusion among investigators, it seems an inevitable outgrowth of the existing organizational structure and may not be a serious enough problem to require major organizational change.

**BOX 4-2**
**THOUGHTS ON HOW TO BALANCE THE ANS PORTFOLIO**

"Have a full-time, permanent program officer who can ensure that the program remains balanced."

"Need to have panelists that will represent the various disciplines and will pay heed to reviewer comments."

"ANS is understaffed; needs as least one or two more program managers, one of whom should be permanent."

"It is ridiculous to manage the diversity of projects presently funded by ANS through a single program manager."

"Continued objectivity of the program manager and assistant manager, who in the recent past have made sincere efforts to learn, seek opinions, and strive for excellence across disciplinary boundaries."

## IMPROVING PROGRAM MANAGEMENT

The question of how to manage such a diverse program and allocate funds equitably is important, and is closely related to the issue of establishing a fair and careful peer review process. The committee brainstormed a number of possible approaches for helping program staff divide the ANS portfolio into more manageable pieces, some based on discipline such as paralleling the structure used within the Antarctic Section and others based on problems to be addressed or environments. In the end, considering the ANS mission and selection guidelines outlined earlier and acknowledging practical considerations (e.g., the number of staff we might realistically expect to be assigned to the program), the committee recommends that proposals be sorted into and evaluated in three "spheres": atmospheric systems, terrestrial systems, and marine systems.

- *Atmospheric systems* would include studies of the troposphere, stratosphere, ionosphere, and space physics, including physical and chemical studies ranging from the Earth surface to the sun, surface exchange processes, and interactions between the biosphere and atmosphere.
- *Terrestrial systems* would include studies of terrestrial biology and ecology, glaciology (including glaciers, ice sheets, snow, and permafrost), land-based geology, Earth surface processes, past environmental history, and freshwater.
- *Marine systems* would include studies of biological and marine ecosystems, marine geology and geophysics, physical and chemical oceanography, sea-ice, and paleomarine issues.

These three spheres are not mutually exclusive and thus a proposal may very well be evaluated and even funded by more than one sphere, or assigned to one sphere for management ease. The value of dividing the ANS program research among three spheres is that this kind of "systems approach" provides both good coverage of the relevant sciences and useful flexibility. Although not all of the disciplines grouped in each sphere share the same intellectual traditions or experimental approaches, they have a commonality of environment/location and share a number of common features and processes. The terrestrial systems sphere is particularly diverse, but still less so than the current ANS program; it could be subdivided again after a trial period if it proves to be too diverse. There will certainly be interaction between and among the spheres, and proposals addressing such interaction would be welcome and placed wherever was most logical. The committee believes the three spheres blend intellectual thrusts in a progressive way they will be more effective in supporting the ANS mission than a traditional disciplinary structure.

The method now used to select proposals to be funded involves use of the peer review system to evaluate proposals, followed by funding decisions made by the program manager (with, recently, assistance from other ANS staff). The main problem with the current process, however, is that one program manager has a large burden of work, both in terms of numbers of proposals and amount of funding to be awarded, and in terms of the wide diversity of the proposals being considered. It is difficult for any one person to make informed, consistent decisions covering the range of topics that typify the ANS program.

As this study was ongoing, some management changes were imposed (i.e., when the committee was formed, ANS was supported by one staff person who was a temporary employee; around the time of the committee's first meeting, another temporary staff person was assigned to provide half-time support to the program; as this report was being written, reviewed, and finalized, additional staff changes were announced).[3] In general, these were positive changes because they increased the amount of staff time allocated to managing the program and set the stage for some permanence in the staff. Still, the committee believes it is appropriate to lay out a potential model of operations that, over the long term, could ensure that the program gives adequate attention to all its elements while increasing its capability to select research that is both of high quality and high importance.

This model is based on the assumption that ANS is large and complex and, in the committee's opinion, would best be served by a team of three staff (e.g.,

[3]As of July 1998, the ANS program has three full-time staff, including its original temporary position and two permanent positions transferred in from the Antarctic Sciences Section, but this is likely to change. The personnel changes that occurred during this study are indicative of the fact that OPP management is aware of the need for change and seeking solutions.

program managers), with one person assigned to manage each of the atmosphere, terrestrial, and marine spheres. The work loads would not be identical, given that (if past patterns hold) it is likely the atmospheric sciences sphere will receive relatively fewer proposals each year. Thus, there will need to be some flexibility in the staffing arrangement.

Together, the three staff members would run the ANS program, coordinating on all major tasks—preparation of announcements, management of incoming proposals, mail-in review, panel review, and funding decisions. When possible, the ANS program managers should take better advantage of the wide variety of expertise that exists elsewhere in OPP and across NSF as a whole. A key addition to the proposal evaluation process would be the use of a multidisciplinary panel convened specifically to consider the ANS proposals from both a quality perspective and for input on research priorities.

Dividing the research funding among the three spheres will require some strategic thinking. Using data provided by OPP, the committee estimated 1997 and 1998 funding shares for the three spheres (had they existed) as shown in Table 4-1. These two sets of numbers indicate, in a rough way, the current proposal pressure in each sphere. Some combination of past funding patterns and proposal pressure could be used to determine an initial division of resources across the spheres. However, given the newness of the program it is also expected that variations from these early patterns are inevitable. While some useful information about funding and proposal submissions by sphere could be gained by examining ANS program history, the division of resources among the three spheres will probably be made based on the best judgment of OPP management. Regardless of how funds are split among the three spheres, the program must be flexible to accommodate complex, innovative multidisciplinary proposals as long as they are of the highest quality and are appropriate based on the selection guidelines. The ANS program should also be open to opportunities to encourage U.S. participation in international efforts because most of the Arctic lies outside the boundaries of the United States, and thus most arctic research includes a significant international component. Major international efforts often have large

TABLE 4-1 Approximate Percentage of 1997 and 1998 Funding Allocated to Areas Covered by the Proposed Three Spheres

| Sphere | % 1997 Funding | % 1998 Proposal Submissions |
|---|---|---|
| Atmospheric systems | 16 | 15 |
| Terrestrial systems | 47 | 42 |
| Marine systems | 34 | 43 |
| Miscellaneous | 3 | 0 |

**BOX 4-3**
**THOUGHTS ON ANS RESEARCH PRIORITIES**

"Research priorities should be determined and implemented through the peer-review process only."

"I would limit the use of themes invented by small numbers of individuals. The strength of NSF is the support to projects of high scientific merit which are initiated by the scientific community. One does a disservice to the scientific community by continually torquing them to refocus or characterize their research in terms of the 'theme of the year.'"

"But even the best science plan won't anticipate new ideas, and room should be left for new ideas that may lead to important insights and breakthroughs, i.e., some risks should be taken."

"I don't think there should be priorities; we have ARCSS for that kind of program. The best science should be funded, with some balance between the fields."

"Setting research priorities would make ANS redundant with other NSF programs, and there would still be little to distinguish the criteria used to choose where to submit proposals."

"Solicit input from the overall science community, form advisory committees and oversight boards, in concert with guidance and interchange with federal science councils and Congress."

scopes, much larger than any single proposal, and are likely to transcend the spheres. It will take some special staff effort to stay aware of these activities.

## SETTING RESEARCH PRIORITIES

Most research support programs use two things to guide their direction: a clear statement of mission and some sense of research priorities. *Mission* is the long-term vision that guides the program; it sets the overall course. Research priorities are dynamic and shorter-term; they guide year-to-year activities and allow programs to respond to newly identified needs, take advantage of special opportunities, and build upon past research in an integrated fashion. Research priorities change and are often quite focused; mission should remain relatively stable and have a broad focus. Mission is best established in a top-down fashion by program and agency management to keep the specific program in line with overall strategic planning and goals. Research priorities can be set either top-down or bottom-up, as long as the mechanisms for doing so are dynamic so that priorities remain current. Research priorities should be written to encompass a range of questions. By definition, research priorities are time dependent—that is, they will change over time. Thus they should always be presented with the caveat that they represent the best judgment at a given moment in time.

There are many possible approaches for setting research priorities. All require some decisionmaker(s) to impose judgment regarding the relative importance of various issues. At times—for example, when new fields of study are emerging, at a significant anniversary or other benchmark, or during a time of great controversy in a field—it can be useful to get a one-time statement of research priorities from an expert committee specifically charged to do so. Such guidance can set research on a sound course and unify a community. But in many cases, it is more important to set up a mechanism to ensure ongoing input from the community, rather than just a one-time glimpse of possible priorities.

What seems to make the ANS program different from many other NSF programs is its breadth — how can one decisionmaker, the program manager, know all the relevant fields equally well and be truly aware of emerging issues and needs? It will always be more difficult to make judgments in fields beyond one's own, although that can be alleviated to some extent by selecting staff with broad backgrounds and over time as the staff's knowledge increases. So the goal is to add elements to the decisionmaking process to, in essence, enlarge the program manager's view.

The three levels of selection guidelines, described earlier, serve to clarify the mission of the ANS program and set broad direction. Some caution is required, however, when talking about research priorities for the ANS program. ANS's fundamental strength is its status as a general program covering a broad range of topics: it is not theme-focused nor should it be. Thus, while program managers will of course have to make decisions about what research is "better" or "of higher priority" than other proposed research, the committee believes that this should not be exaggerated to the point where the priorities take on so much weight that they become "themes" dictating what research is sought out and selected.

We further believe that the review panels, Arctic Affiliates, and other existing sources of expertise can play roles in helping ANS managers identify areas of important research and maintain a dynamic balance between the needs in the three scientific spheres. Priority setting should not be a one-time event but a flexible process, complete with periodic evaluations and modifications. The program can remain flexible and the appropriate home for an eclectic assortment of projects, while still listening to the scientific community about what types of research are of high importance as time progresses.

## LOGISTICS SUPPORT

At the town meetings run by the committee and through the electronic questionnaire posted on the Polar Research Board's Web site, the research community

**BOX 4-4**
**THOUGHTS ON LOGISTICS**

"Because of the international cooperation of my projects, the amount needed is much lower than would be encountered if NSF had to supply all the logistical support. As a result, my logistics would be on the order of only about 10 percent of my grants. Without international cooperation, this could easily be up around 50 percent."

"ANS should give consideration to helping leverage funds for principal investigators when logistics costs are covered."

"Work in Alaska in particular would be greatly enhanced by agreements and memoranda of understanding for shared logistical resources through the U.S. Fish and Wildlife Service, National Park Service, Bureau of Land Management, and Native corporations."

"Improve coordination of arctic research platforms. Since they cost so much to operate/maintain, we should make sure that each and every one is exploited to maximum efficiency."

expressed concern about issues related to support for the logistics associated with research in the Arctic. Many comments related to finding platforms for field work, and there was virtually universal concern that logistics costs for arctic work had to be borne by science budgets. This is a disadvantage relative to work in the Antarctic where the logistics costs are covered separately by the Polar Research Support Section. Therefore, even though the science budgets of the Arctic and Antarctic Sections appear equal size, a significantly smaller share of the Arctic budget is actually available for science. (For example, a $100,000 grant from a program in the Arctic Sciences Section might cover $25,000 in logistical expenses like ship time or charter air services and $75,000 in research expenses; in the Antarctic Sciences Section, all $100,000 would support research, while transportation costs would be covered separately.) The committee recognizes that comparisons between arctic and antarctic logistics are not entirely apt. For instance, access to research sites in the Arctic is not as difficult in many ways because extensive infrastructure is already in place—communities with facilities, airfields, and shipping points that can serve as staging bases for remote field operations. The capability exists to reach almost any location in the Arctic by aircraft during at least some seasons of the year, and the capability of sampling year-around has been demonstrated numerous times by Russia, the United States, and Canada. By contrast, NSF must supply all the infrastructure for U.S. antarctic research and year-around access. Nor are arctic researchers eager to deal with a large bureaucracy and standardized procedures as is necessary in the Antarctic, because great flexibility is often necessary in arranging logistics in the Arctic. Nevertheless, it seems unrealistic to expect ANS and the rest of the Arctic Sciences Section to take leading roles in arctic research without devoting more attention to the associated logistical challenges.

Concerns about logistics, access, and platform availability are discussed at length in other reports (e.g., see Schlosser et al., 1997). In the past, one of the most prevalent complaints related to logistics support for arctic research has been the lack of a dedicated research icebreaker. The imminent commissioning of the USCGC Healy is changing this concern. Now, many scientists wonder whether already-constrained financial resources will be stretched to pay for all the ship time that will be available. Terrestrial, marine, and atmospheric scientists who do not require an icebreaker to do their work are worried that a great deal of ANS funding will be siphoned away for ship support.

The cooperative arrangement with the Navy that facilitates the use of Navy submarines for research and the developing arrangement with the Coast Guard are examples of how collaboration can improve access to arctic research sites. Because a number of federal agencies have roles in the Arctic, certainly there are other opportunities to share resources—ships, aircraft, shore stations, and technology—in ways that give researchers logistical support. Potential partners who might share logistics resources include the Forest Service, National Oceanic and Atmospheric Administration, National Park Service, and various Alaska state agencies. It is not clear if these opportunities are being fully explored by NSF.

Finally, logistical support for research in the Arctic faces the sometimes complex problem of access to foreign countries and territorial waters. This is a political and bureaucratic problem, and it is very difficult for principal investigators to know all the relevant laws and procedures and make the necessary contacts to conduct experiments in the economic zone of another country. Shipping and customs in foreign countries can be extremely difficult without the proper contacts, even when the work is done with a collaborator in the foreign country. It is not realistic to expect the ANS program managers to juggle their scientific responsibilities and deal with anything more than the most routine logistical questions as well.

# 5

# Conclusions and Recommendations

The Committee on Science Priorities for NSF's Arctic Natural Sciences Program was charged with reviewing the program's management and research strategy and providing guidance on how to set research priorities given the diverse scientific issues that fall within its purview. In its study, the committee found that the Arctic Natural Sciences (ANS) program considers proposals in an exceptionally wide range of fields and that this poses significant management challenges. Some of these challenges are typical of a program that is new and still evolving, while others relate to the program's size and scope.

The committee believes the ANS program makes important contributions to research in the Arctic, but there are opportunities for improvement. ANS staff should find ways to get help in judging proposals through collaboration with other National Science Foundation (NSF) personnel and by getting input from review panels established specifically to consider proposals from a multidisciplinary perspective. Better internal communications are prerequisites for a unified Arctic Section. And increased outreach efforts to other agencies and programs are needed to build relationships that, over time, could make the program a more effective participant in interagency and international collaborations. Finally, because limited budgets mean that decisions must be made about which research is "best" or "most important," ANS staff can get input on which research areas care important through the peer-review process, interaction with the arctic research community, and planning other activities as needed.

This chapter provides the committee's conclusions and recommendations on how best to implement these desired improvements. The conclusions state the main lessons the committee drew from its evaluation; the recommendations include both general guidance and some detailed prescriptions for strengthening the ANS program. The committee took the present Office of Polar Programs (OPP) organizational structure as a boundary condition and, although members brainstormed more radical approaches, our conclusions are based on the following three assumptions: (1) The ANS program will continue to exist as a broadly defined subsection of the OPP's Arctic Sciences Section, with equivalent status to the Arctic System Science (ARCSS) and Arctic Social Sciences (ASSP) programs. (2) The definition of the areas of science supported by ANS is largely as stated in its request for proposals (atmospheric sciences, biological sciences, earth sciences, glaciology, and oceanography). (3) Funding for science within the Arctic Sciences Section and allocations among programs will continue without major change.

## PROGRAM SCOPE AND STRUCTURE

There is a significant need for a research support program with the diverse focus of the ANS program. Despite some inherent overlap with other NSF programs, the committee believes that ANS should be the central program within the National Science Foundation covering research in the natural sciences in the Arctic.

### Conclusions

- Opportunities for funding within ANS should continue to be unfettered by emphasis on thematic considerations. Individual as well as joint or cooperative proposals should have equal access to the available funding.
- A substantial advantage of the present ANS structure is that it can accommodate and encourage the growing trend toward multidisciplinary research because it is not confined by traditional disciplinary boundaries. The present broad scope of the program should continue.
- The sometimes unclear boundaries between the ANS program and other NSF programs with arctic elements create a management challenge that is best addressed soon, while the program is still young and malleable. The solution lies primarily in improved management rather than in restructuring or redefining the program.

### Recommendations

- To best reflect the broad purpose of the ANS program within the OPP context, OPP should adopt the following mission statement for the ANS program: "The mission of the Arctic Natural Sciences program is to fund cutting-edge research dealing with any aspect of the Arctic's atmospheric, terrestrial, and marine systems. The program focuses on proposals that contribute to understanding and predicting the unique elements and processes that are part of the arctic environment."

- ANS program staff should use the guidelines provided in Chapter 4 or some other set of clear guidelines to help make judgments about whether proposed research fits within the ANS mission and should publicize the guidelines to help reduce confusion about overlap with other NSF programs.
- The head of the Arctic Sciences Section should divide the ANS's broad scientific program into three spheres, each of which is larger than a single discipline. The suggested spheres are: atmospheric systems, terrestrial systems, and marine systems. These spheres should have flexible boundaries in terms of both subject matter and allocation of dollars; in general, funding would follow current patterns. The spheres would be defined as follows:

> *Atmospheric systems* would include studies of the troposphere, stratosphere, ionosphere, and space physics, with the latter including physical and chemical studies from the Earth's surface to the sun, surface exchange processes, and interactions between the biosphere and atmosphere.
>
> *Terrestrial systems* would include studies of terrestrial biology and ecology; glaciology (including glaciers, ice sheets, snow, and permafrost); land-based geology; Earth surface processes; past environmental history; and freshwater.
>
> *Marine systems* would include studies of biological and marine ecosystems, marine geology and geophysics, physical and chemical oceanography, sea-ice, and paleomarine issues.

These three scientific spheres are not mutually exclusive; proposals may overlap spheres or deal with interactions among spheres. Consequently, a proposal may very well be evaluated and funded by more than one sphere, or assigned to a single sphere for easier management. The value of dividing ANS research among three spheres is that this structure blends related intellectual thrusts while imposing a more manageable limit on the diversity and volume of proposals to be considered and compared.

- Program managers and staff should treat the spheres as of intrinsically equal importance, although they do not have to be equal in terms of the number of proposals submitted or funded, or in allocation of dollars. (Related recommendations dealing with proposal review and funding allocation under the three sphere scenario appear later in this chapter.)

## MANAGEMENT STRATEGY

The committee reviewed the ANS program's current management strategy to determine whether its strategy and staffing levels are adequate. Since the committee began its work, some management changes have occurred. When we began, the program was staffed by one staff person; other staff—some full time and some part time—have since been assigned to assist, and some permanent personnel have been transferred to the program. These are steps in the right direction, but because they were ongoing as this report was being prepared, we provide the following thoughts with the original management structure as our base.

### Conclusion

- The diversity of subjects covered by the ANS program is too broad for a single program manager to cover with adequate depth and consistency, and adjustments in management strategy and staffing are needed to ensure proper review of proposals and balanced decisionmaking about funding.

### Recommendations

- To improve the management of the ANS program, OPP could assign management responsibility to be shared among three staff members, one responsible for each sphere. These staff should be of equal status and report to the Arctic Section Head. Care would need to be taken not to create three "mini-programs," which might lose the benefits of size and multidisciplinary focus that are ANS's chief benefits, and which also might be eclipsed by larger programs like ARCSS.
- The challenge of managing a diverse scientific program requires staff with particularly broad backgrounds, because staff must be able to identify the best and most innovative research proposals within all the fields encompassed by the program. For this reason, the program manager position(s) should not be filled with rotating employees who serve for only two years or similarly short periods. Alternatively, if there are multiple managers (as suggested above) at least one or two should be permanent. The argument for making the program managers permanent is based on the committee's belief that there are few working scientists who could adequately and quickly become effective in this broadly-based program, and it will always be difficult to replace incumbents at the end of any two-year term. Gaps in staffing would be problematic.
- The Arctic Section Head provides leadership to the entire Arctic Section, but because of the key role and diversity of the ANS program he or she should serve more directly as the lead supervisor overseeing the ANS program staff and continually reassessing the balance, interactions, cooperation, and resource partitioning among the spheres. If the Arctic Section head position remains one filled by rotating employees, it must continue to be someone of significant authority.
- OPP should make a clear statement to the scientific community updating researchers about the mission of the ANS program and its relationships with

other programs. Such a statement could also clarify logistical support issues that seem perplexing to many researchers (discussed in more detail later in this chapter).

## SETTING RESEARCH PRIORITIES

What seems to make management of the ANS program challenging is its breadth—how can one decisionmaker, a program manager, know all the relevant fields equally well and truly be aware of emerging issues and needs? It will always be more difficult to make judgments in fields beyond your own, although that can be alleviated to some extent by selecting staff with broad backgrounds and over time as the staff's knowledge increases. Using the three spheres should reduce the need to compare across disciplinary boundaries, and additional staff will increase in-house expertise available to make judgments. But other mechanisms can be incorporated into the decision-making process to, in essence, enlarge the program manager's view and help identify high-priority research projects.

Setting priorities should involve input from at least three groups: *NSF management,* to be sure that the priorities selected support broader agency goals and strategic planning; the *scientific communities who will be requesting funds,* to be sure that the priorities selected represent what they believe are the most important and cutting-edge issues; and, finally, *representatives of related research programs and agencies,* to be sure there is coordination of effort and to limit duplication of effort. This input can be gained in a one-time exercise like a workshop, a more long-term approach would be to use existing activities to solicit ongoing input—that is, use existing mail review and panel review processes to judge proposal quality and gain insight into the level of importance of the work. Other mechanisms to assist in priority-setting can be using a Committee of Visitors tailored to give appropriate input, involving subcommittees of the Office Advisory Committee, and hosting town meetings at large conferences such as the annual meeting of the American Geophysical Union. These processes, combined with input from NSF staff and input from other agency staff, should be adequate to reflect the full range of views. It is the job of the program staff to synthesize the information and make final judgments about priorities.

If the ANS program elects to impose the three spheres suggested by this committee, it will help with the process of identifying research priorities on an ongoing basis because the mail reviews will be focused by sphere. The panel reviewers, with some additional instruction, can then help balance the total portfolio. Using the three spheres should reduce the necessity to compare "across" disciplinary boundaries. The additional staff we envision associated with the three spheres, in turn, increases the in-house expertise available to make judgments.

## Conclusions

- The fundamental strength of the ANS program is that it is a general program covering a broad range of topics. Priority-setting will always be a part of decisionmaking where budgets are limited, but the selection of specific themes that dictate program direction and are solicited as such should not be the driver behind the ANS program.
- The setting of priorities should not be a one-time event but a flexible process. Mail-in reviews, panel reviews, and other NSF and agency staff with arctic expertise can play key roles in helping ANS management identify priorities and maintain a dynamic balance among the needs in the three research spheres. Mail-in and panel reviews are especially important because they foster involvement of active researchers and allow access to the community's thinking about what areas are important for investigation.

## Recommendations

- The first stage in proposal review is to separate proposals into the three research spheres with all staff participating in the assignment of all proposals. Mail-in reviews would then be sought via standard OPP procedures from carefully selected scientists, by sphere. After the mail-in reviews are received and considered, each ANS staff member would select the top proposals in his or her sphere (e.g., the top 30 percent, or whatever proportion seems appropriate given the proposal pressure and funding available) to be considered as serious candidates for funding.
- Next, ANS staff should, under the leadership of the Arctic Section Head, set up a multidisciplinary review panel to meet and consider the top proposals. The panel should include two members of the research community from each sphere plus representatives of the Arctic Affiliates (that is, all NSF program managers whose portfolios include arctic research projects and selected representatives from other federal agencies supporting arctic research). At this meeting, the top ANS proposals should be considered in the context of the totality of known research activity in the upcoming year. The proposals should be discussed individually and priority for funding suggested, in accordance with typical NSF panel review procedures.
- The proposal selection process within ANS could be improved by involving non-OPP program managers who fund studies related to arctic natural science in the panel reviews. Finding ways to involve representatives of other federal agencies that fund or conduct related research in the Arctic would also, over time, promote the efficient use and leveraging of research resources; help reduce duplication; and increase collaboration to support common facilities, logistics, and research goals. Better coordination with other units of NSF might be facilitated by appointing liaisons from the most appropriate units (e.g., the Geosciences or

Biology Directorates) to attend meetings of the OPP's Office Advisory Committee and vice versa.

- Although there should be broad input into this process—including the mail-in reviews, the advice of the Arctic Affiliates, and advice from the multidisciplinary panel—ultimate decisionmaking should remain with the ANS program staff. It is very important that the final decisions be an agreed-upon compromise between competing viewpoints and science agendas. The Arctic Section Head should help to broker compromise as needed among program staff and the three spheres.
- If OPP decides that a list of current research priorities is needed to guide the ANS program as it leaves its start-up phase and matures, ANS managers or an outside body can be charged to convene a workshop involving principal investigators, other researchers interested in the areas funded by ANS, and the Arctic Affiliates, including representatives of the ARCSS program so as to clarify and maintain the distinctions between ARCSS and ANS. The objectives of such a workshop would be:

> — to give the ANS program managers and the head of the Arctic Section direct input from the research community regarding their sense of future research goals within each of the three spheres and the links between them; and
>
> — to discuss the current range of ad hoc international research activities and connections that exist within the research community and how these links might be formalized and strengthened.

A workshop might also explore the ongoing issue of logistics support and, in particular, the potential impacts of the new Coast Guard ship, USCGC Healy, on the ANS program. Such a workshop should be held only if OPP determines that it needs a list of concrete research priorities to steer the ANS program, because additional meetings require time from researchers and use funds that could otherwise be applied to support research. Also, care would be necessary to work against the tendency for meetings to lead to "big science" and to be sure that the ANS door remains open to individual researchers. The greatest challenge in organizing a workshop would be to maintain the interest of the diverse participants.

## AGENCY AND INTERNATIONAL COOPERATION

Since many nations have territory in the Arctic and even more are actively involved in arctic research, international cooperation is critical in optimizing opportunities and cost effectiveness in the pursuit of research. Building international cooperation is a long-term activity, and so it is not surprising that it is not yet a large element of the ANS program. As staff numbers and experience increase, it will be easier to develop the necessary connections to and knowledge of international activities.

Similarly, several federal agencies have responsibilities that lead them to be involved in research in the Arctic. Cooperation is critical so these efforts are not duplicative and are coordinated to best advantage. Coordination and communication within units of the National Science Foundation are, of course, also essential since a number of programs have arctic natural science dimensions. A coordinated U.S. effort is vital to ensure that the nation participates fully and in a unified manner in international activities.

## Conclusions

- International cooperation and collaboration should be fostered and encouraged. Although a research proposal cannot be judged solely on whether it has an international dimension, appropriate collaborations across national boundaries should be considered an advantage. Such collaborations can leverage resources, open access to data, and ensure that the United States has input to, and receives contributions and in-kind support from, international efforts.
- Interagency cooperation and collaboration are critical to efficient use of limited resources.
- Communication and collaboration among the NSF units that address various aspects of natural science in polar regions needs to be improved, including between the Antarctic and Arctic Science Sections and between the ANS program and relevant programs outside of the Office of Polar Programs.

## Recommendations

- ANS program staff can take steps to encourage participation in international scientific investigations. For instance, ANS staff should be intimately familiar with international opportunities to leverage resources or to capitalize on the availability of facilities or research platforms. Such opportunities often emerge from international research initiatives; staff should promote the necessary linkages.
- ANS staff should facilitate and give careful consideration to research that proposes to gather data of use to the international arctic science community and that involves collaboration with international partners. The program should be open to funding travel grants or supplements, within the existing proposal selection process, so participants can meet for international planning purposes.
- Timing is often an important factor in collaborative international efforts. Projects developed among many nations often have a long planning horizon, and U.S. participants must work within that schedule. Where possible, ANS should be flexible in the timing of awards and provision of logistics support to match the needs of international projects.

- NSF's International Programs Division should be asked to assist financially with the development of international scientific collaborations.
- The Interagency Arctic Research Policy Committee already exists as a mechanism to help agencies coordinate their arctic research activities, and the community of involved federal agencies is not so large that another formal mechanism is needed. But improved communication and increased informal interaction among key staff are needed in order to establish a foundation for real cooperation—including, perhaps, involvement of appropriate agency personnel as Arctic Affiliates involved in aspects of the review process.
- ANS program staff should take steps to improve communication with other NSF units with relevant programs, such as the Earth System History program. Benefit can be gained by increased informal interactions among staff, but also via mechanisms such as periodic meetings of the proposed Arctic Affiliates or invitation of appropriate staff to meetings of the Office Advisory Committee. Sharing expertise in these ways can do much to bring a wide range of expertise to the ANS program and help its staff manage the program's diverse portfolio.

## LOGISTICS SUPPORT

The committee heard numerous expressions of concern from the research community about issues related to logistics support for research in the Arctic. Many comments related to finding platforms for field work, and there was virtually universal concern that logistics costs for arctic work had to be borne by science budgets. The committee recognizes that comparisons between arctic and antarctic logistics are not entirely apt. For instance, access to research sites in the Arctic is not as difficult in many ways because of the extensive infrastructure that is already in place. Nor are arctic researchers seeking to create a large bureaucracy or standardized procedures, because great flexibility is often necessary in arranging logistics in the Arctic. Nevertheless, it seems unrealistic to expect ANS and the rest of the Arctic Section to take leading roles in arctic research without devoting more time to the associated logistical challenges.

### Conclusion

- OPP must provide significant personnel and funds to support logistics needs in the Arctic. This support need not be "equal" to the antarctic logistics support provided by the Polar Support Section, because the settings are very different. In fact, many in the arctic research community appreciate the flexibility they have to arrange their own logistics. But there is extensive community dissatisfaction with the current approach.

**Recommendation**

- An office dedicated to arctic logistics could be incorporated into the Arctic Section or into the Polar Support Section. The office's role would be mainly one of coordination, although it could take on more responsibility where scientifically and economically beneficial, and take a less active role where the principal investigator has the knowledge or access to be more effective. Generally, an office for arctic logistics would:

— Know the predicted logistics needs of funded proposals for at least the next two years, and help in coordinating activities and providing the greatest support at the least cost. The office could track opportunities for researchers to share logistics capabilities.

— Know what governmental and commercial facilities are available for providing logistics support and help investigators take advantage of this support. Besides acting as a liaison with the Coast Guard and Navy for ship support, the office could know what other science aircraft and support are available from the Department of Defense, Forest Service, National Oceanic and Atmospheric Administration, National Aeronautics and Space Administration, and other branches of the National Science Foundation. The office would try to arrange for use of these platforms at the least possible cost.

— Become familiar with the procedures and contacts needed to work in other arctic nations and their territorial waters. These requirements change with time and vary for different research platforms. Sometimes, gaining approvals for work in other countries is easier if an overarching agreement between national agencies is already in place. In these areas, the arctic logistics office could advise NSF and other federal agencies of the need for such agreements and help in their implementation. In addition, the office would also have to gauge when its participation would be less effective or even counterproductive compared to personal contacts by individual scientists.

## PROGRAM DATA AVAILABILITY

The ANS program is relatively young, and thus not much information about it has been accumulated. In the course of its work, the committee requested certain data describing ANS awards and funding patterns that were needed to support this assessment. The requested information was not readily available and compiling it required a significant amount of work. An OPP Committee of Visitors had similar difficulties in getting information, which it attributed to OPP's reorganization, inadequate computing capabilities, and a lack of thought about what types of data should be recorded. In addition to needing information for program planning, accurate accounting of the ANS program's accomplishments will be necessary within the National Science Foundation's overall effort to comply with the Government Performance and Results Act of 1993, which seeks to improve the effectiveness of federal programs through strategic planning, goal setting, and performance assessment.

### Conclusion

- Good record-keeping is essential to good program management. Accurate, basic information on the size and types of awards and other standard information should be kept in computer databases that are easily accessible to staff and easily interpreted.

### Recommendation

- If it does not already exist, OPP should develop an organized, workable database containing appropriate proposal information so it can be an accessible and useful resource to staff, supervisors, and others to monitor the ANS program as it evolves. If such a system already exists, ANS program data should be entered so that over time a useful body of information is accumulated that will hold lessons about program strengths and weaknesses and guide future decisionmakers.

# References

Interagency Arctic Research Policy Committee (IARPC). 1997. U.S. Arctic Research Plan: Biennial Revision: 1998-2002. In Arctic Research of the United States, Volume 11.

National Research Council (NRC). 1977. An evaluation of Arctic programs supported by the National Science Foundation. Washington, D.C.: National Academy Press.

National Research Council (NRC). 1998. Results of Polar Research Board electronic questionnaire. Unpublished.

National Science Foundation (NSF). 1997a. Background information provided to the Committee on NSF's Arctic Natural Science Program. Memorandum.

National Science Foundation (NSF). 1997b. Report of 1997 Office of Polar Programs Committee of Visitors, June 23-June 26, 1997. Memorandum.

National Science Foundation (NSF). 1998. Arctic Research Program Opportunities — Annual Target Dates: February 15 and August 1. Office of Polar Programs. Arlington, VA: National Science Foundation.

Schlosser, P., W. Tucker, N. Flanders, and W. Warnick, eds. 1997. Logistics recommendations for an improved U.S. Arctic research capability. Fairbanks, AK: The Arctic Research Consortium of the United States. 88 pp.

# APPENDIX A

# Arctic Research Opportunities

## INTRODUCTION

The National Science Foundation (NSF) invites U.S. scientists to submit proposals for research in all of the Arctic and to perform arctic research based at institutions in the United States.

The goal of the NSF Arctic Research Program is to gain a better understanding of the Earth's biological, geological, chemical, and socio-cultural processes, and the interactions of ocean, land, atmosphere, biological, and human systems. Arctic research is supported at NSF by the Office of Polar Programs (OPP) (http://www.nsf.gov/od/opp), as well as by a number of other disciplinary programs within the Foundation that are linked through an internal NSF Arctic Affiliates system. This system, consisting of program representatives from other NSF programs that support arctic research, provides coordination across NSF, including a structure that enables joint review and funding of arctic proposals, as well as mutual support of special projects with high logistical costs.

OPP offers focused multidisciplinary and interdisciplinary programs that emphasize the uniqueness of the Arctic for special scientific studies. Models indicate that the arctic regions are among the most sensitive to environmental change. They have a long natural climate record and thousands of years of human

This Appendix is extracted from National Science Foundation Document nsf 9872, which may be found in its entirety on the NSF's Web site at http://www.nsf.gov.

settlement. This interplay provides a unique basis for integrated research on global systems and human adaptation.

OPP disciplinary interests encompass the atmospheric, biological, earth, ocean, and social sciences. Interdisciplinary research in the biosciences, geosciences, and social sciences is linked in the Arctic System Science Program. In addition to supporting research on long-term human-environment interactions, OPP encourages the study of contemporary socio-economic, cultural, and demographic issues in the changing political environment of the post-Cold War world. The OPP also encourages bipolar research, especially glaciology, permafrost, sea ice, ecology, and aeronomy. Increasing emphasis is being given to the integration of research and education. Scientific programs connected to students (K-12 and above), affected communities in the north, and the general public's improved understanding of basic research are strongly encouraged. Educational components are encouraged with proposed research in all disciplines and programs, but stand-alone proposals will also be entertained.

The United States Arctic Research and Policy Act of 1984 defines the Arctic as all areas north of the Arctic Circle and all United States territory north and west of the boundary formed by the Porcupine, Yukon, and Kuskokwim Rivers; all contiguous seas including the Arctic Ocean and the Beaufort, Bering, and Chukchi Seas, and the Aleutian chain. Field projects falling outside these boundaries but directly related to arctic science and engineering conditions or issues, such as laboratory and theoretical studies, are appropriate.

The Foundation is one of twelve Federal agencies that sponsor or conduct arctic science, engineering, and related activities. As mandated by the Arctic Research and Policy Act of 1984, Federal interagency research planning is coordinated through the Interagency Arctic Research Policy Committee (IARPC) which is chaired by NSF.

Under an agreement with the U.S. Navy, access to a nuclear submarine is available annually for research in the Arctic Ocean. Researchers are strongly encouraged to pursue this possibility with OPP or directly with Office of Naval Research (ONR). Further information on other agency programs is presented in the journal Arctic Research of the United States (NSF 96-130), and the U.S. Arctic Research Plan and its biennial revisions (NSF 97-148).

As the Arctic is the homeland of numerous Native peoples, special attention must be given to all aspects of research and education that may potentially impact their lives. An interagency statement of "Principles for the Conduct of Research in the Arctic" has been developed and all arctic research grantees are expected to abide by these guidelines. These guidelines are presented in the appendix section of this program announcement.

In fiscal year 1997, NSF supported 362 Arctic research projects for a total of $49.39 million. Of this, $30.71 million was from the OPP Arctic Research Program.

A compilation of all NSF arctic and related research grants for each fiscal year is available (NSF 97-78 or www.nsf.gov/cgi-bin/getpub?nsf9778 for FY 1996). The current NSF Guide to Programs (NSF 97-150 or www.nsf.gov/cgi-bin/getpub?gp) should be consulted for additional program information.

## RESEARCH PROGRAMS

Listed below are the principal OPP programs that support arctic research. There are three integrated programs in OPP: Arctic Natural Sciences, Arctic Social Sciences, and Arctic System Science. Support is also provided for data and information management research activities. These programs and their components are described below.

### Arctic Natural Sciences Program

The OPP Arctic Natural Sciences Program supports research in glaciology and in the atmospheric, biological, earth, and ocean sciences. This program provides core support for disciplinary research in the Arctic and coordinates its support of arctic research with the Directorates for Geosciences and Biological Sciences. Areas of special interest include: marine and terrestrial ecosystems, atmospheric chemistry, exploration of the Arctic Ocean, as well as Arctic geological and glaciological processes.

#### *Atmospheric Sciences*

Research in arctic atmospheric sciences focuses on stratospheric and tropospheric processes as well as arctic climate and meteorology. Research on past climates and atmospheric gases as preserved in snow and ice cores have also been supported as has research on atmosphere-sea and atmosphere-ice interactions.

In upper atmospheric and space physics, research interests include auroral studies, atmospheric dynamics and chemistry as well as magnetosphere-ionosphere coupling. Conjugate studies are considered jointly with the Antarctic Aeronomy and Astrophysics Program.

#### *Biological Sciences*

OPP supports projects that emphasize understanding of the adaptation of organisms to the arctic environment. Biological studies in the Arctic include: research in freshwater, marine, and terrestrial biology; organismal adaptation to the arctic environment; ecology; ecosystem structure and processes; and the biological consequences of ultraviolet radiation. OPP also participates in the Life in Extreme Environments (LExEn) initiative (NSF announcement 97-157).

*Earth Sciences*

Research supported by OPP includes all sub-disciplines of terrestrial and marine geology and geophysics. Special emphasis is placed on understanding geological processes important to the arctic regions and geologic history dominated by those processes.

*Glaciology*

The OPP is the focal point for glaciological research within the Foundation. Glaciological research is concerned with the history and dynamics of all naturally occurring forms of snow and ice, including seasonal snow, glaciers, and the Greenland ice sheet. The Arctic Natural Sciences Program also includes ice dynamics, modeling, glacial geology, and remote sensing studies of ice sheets.

*Ocean Sciences*

The goal of oceanographic research in the Arctic is to develop knowledge of the structure of the Arctic Ocean and adjacent seas, their physical and biological interactions with the global hydrosphere, and the formation and persistence of the arctic sea-ice cover. Areas of special interest are: the distribution of life in high latitude oceans; low temperature life processes; the formation, movement, and mixing of arctic water masses; the growth and decay of sea ice; the exchange of salt and heat with the Atlantic Ocean and the Bering Sea; geographical anomalies; sedimentary history and the role of the Arctic Ocean and adjacent seas in global climate. Proposals concerned with the interdependencies of chemical and physical processes and marine organisms and productivity are encouraged.

## Arctic Social Sciences Program

The OPP Arctic Social Sciences Program encompasses all social sciences supported by NSF. These include anthropology, archaeology, economics, geography, linguistics, political science, psychology, sociology, and related subjects.

Although unsolicited proposals in any of the social sciences mentioned above are welcome, areas of particular interest include: rapid social change (including the processes and consequences of social, economic, and cultural change), community viability (including issues related to community and/or cultural vitality and survival), and human/environment interactions (including issues related to subsistence and sustainable development).

The Arctic Social Sciences Program especially encourages projects that: include indigenous peoples; are circumpolar and/or comparative; integrate social and natural sciences; involve collaborations between researchers and those living in the Arctic; include traditional knowledge; or form connections among disciplines, regions, researchers, communities, and/or students (K-12, undergraduate, or graduate).

Projects involving research with human subjects must ensure that subjects are protected from research risks in conformance with the Common Rule (Federal Policy for the Protection of Human Subjects, 45 CFR §690). Before issuance of an NSF award, all projects involving human subjects must either have approval from the organization's Institutional Review Board (IRB) or identify the applicable subsection exempting the proposal from IRB review, as established in section 101(b) of the Common Rule. Submission of the IRB approval form or indication of exemption should be included in Section I of the proposal (see Grant Proposal Guide NSF 98-2, page 14). Section I should also include letters describing any other permission or approval, such as from Native organizations or communities in which the work will take place.

The Arctic Social Sciences Program considers joint review and funding with other NSF and OPP programs, when appropriate. Special funding opportunities may also be available through NSF's Environment and Global Change activities (see "Crosscutting Areas of Research and Education" in Guide to Programs (NSF 97-150) or the Arctic System Science (ARCSS) Program (refer below).

## Arctic System Science (ARCSS) Program

The ARCSS Program supports interdisciplinary research, whose goal is to 1) understand the physical, geological, chemical, biological and sociocultural processes of the arctic system that interact with the total Earth system and thus contribute to or are influenced by global change, in order to 2) advance the scientific basis for predicting environmental change on a seasonal-to-centuries time scale, and for formulating policy options in response to the anticipated impacts of global change on humans and societal support systems. In order to achieve the goals of ARCSS an emphasis is placed on four scientific thrusts: understanding global and regional impacts of the arctic climate system and its variability; determining the role of the Arctic in global biogeochemical cycling; identifying global change impacts on the structure and stability of arctic ecosystems; and establishing the links between environmental change and human activity.

ARCSS directs most available support to large integrated research projects that are proposed and implemented in response to science plans developed by the scientific community through Science Steering Committees (SSCs) for each component of ARCSS. However, global change proposals from individual investigators or small groups of investigators are also welcome.

ARCSS has three linked components for which proposals are encouraged: 1) Ocean/Atmosphere/Ice Interactions (OAII); 2) Land/Atmosphere/Ice Interactions (LAII); and 3) Paleoenvironmental Studies. The third component has had two projects: Paleoclimates from Arctic Lakes and Estuaries (PALE) and Greenland

Ice Sheet Program Two (GISP2). Paleoenvironmental proposals are now considered within the Earth System History initiative of the United States Global Change Research Program (NSF 97-161). ARCSS further develops a fourth component: Human Dimensions of the Arctic System (HARC). The HARC science plan is accessible through the World Wide Web home page of the Arctic Research Consortium of the US (ARCUS), http://arcus.polarnet.com, and the NSF Web Page should be consulted for new developments. In all these components proposals for new and different research topics are encouraged.

ARCSS also supports the integration of research results across components within ARCSS as well as with any other Arctic research programs through a Synthesis, Integration and Modeling Studies (SIMS) effort. Science plans approved by each SSC and examples of projects supported within each component and SIMS are accessible either on the World Wide Web site maintained by the ARCSS Data Coordination Center of the University of Colorado National Snow and Ice Data Center, http://arcss.colorado.edu/ or the ARCUS home page.

Successful proposals have been funded by the Office of Polar Programs, the Divisions of Atmospheric Sciences and Ocean Sciences within the Directorate for Geosciences, by the Division of Environmental Biology, Directorate for Biological Sciences and, in some cases jointly with ONR, NASA, and DOE.

The Arctic system consists of physical, biological, and cultural factors that may respond to global change. Some models that predict the climatic response to global change show greater change in the Arctic than any other region. The predicted climatology, however, may not consider the largely unknown interannual variability in the Arctic. The presence of cultural institutions in a region subject to possibly large perturbations, however, makes it important that scientists better understand interactions of the global and Arctic systems. Therefore, the research supported in ARCSS extends beyond purely observational studies to those studies that predict and analyze the consequences of global change important to wise stewardship of renewable resources and development of policy options for resource managers and residents.

In order to focus on the Arctic system at a scale that incorporates the multiple environmental feedback mechanisms involved, large interdisciplinary projects that integrate major elements of the system will be supported.

For more information on how a research proposal might best fit the programs and themes of ARCSS, contact the program manager.

## OTHER ARCTIC SUPPORT

### Arctic Research and Policy

OPP supports the management of arctic data and information, including development of the Arctic Data Directory (ADD). The ADD is accessible on the World Wide Web (http://www-ak.wr.usgs.gov/aedd/history.html), and contains information on several hundred arctic data sets. The objective of this type of support is to make arctic data and information more readily available to researchers. Proposals to integrate data and information management are especially encouraged.

### Arctic Logistics

An arctic logistics budget component has been established in the OPP to address field program requirements beyond those commonly included in proposals from an individual or small group of investigators. Examples of the type of logistic and research support that may be provided through the separate logistics component dependent on the site, include: ski-equipped heavy-lift aircraft, research vessels, multi-investigator helicopter or aircraft use, large remote field camps, day-use fees at major Arctic research sites where NSF has established a support structure (currently, Toolik Field Station, Barrow, and Prudhoe Bay/ Deadhorse, AK, Kangerlussuaq and Thule, Greenland), and special transportation needs required for arctic research or available at the sites listed above.

There are two types of logistics support: Type 1) routine requirements typically included in proposals from an individual or small group are dealt with by the PI, and included in the cost budget, and Type 2) coordinated requirements such as those listed above and supplied by an OPP contractor or cooperating agency are arranged in cooperation with the NSF program manager; these costs are not included in the individual investigator's proposal budget. All investigators proposing research projects that may require such support are encouraged to contact the appropriate program manager or the NSF Arctic Logistics Coordination Specialist to determine if they are eligible for Type 1 or Type 2 logistics support. Type 2 support will require completion of the Logistics Coordination Form.

## SPECIAL PROGRAMS

The Guide to Programs (NSF 97-150) provides guidance regarding special funding opportunities including international cooperative activities, human resources development, and other programs. Described below are some of these opportunities that are supported by the OPP Arctic Program. For details about these programs and other programs refer to the Guide to Programs; the NSF Home Page (http://www.nsf.gov), or the NSF publications noted below.

### Small Grants for Exploratory Research (SGERs)

These awards are intended to provide support for small scale exploratory, high-risk research involving preliminary work on untested ideas, ventures into emerging areas, or research having severe urgency, etc. Proposers are strongly encouraged to contact the cognizant program officer before submission.

## Small Business Innovation Research Program (SBIRs)

The Small Business Innovation Research Program (NSF 97-64) primarily facilitates research on advanced concepts in scientific or engineering areas, particularly where the research may serve as a base for technological innovation. Polar topics of interest include cold-weather design, remotely operated and autonomous vehicles and sensors, geophysical and space technologies, biotechnology, and long-term operations and construction. Technical projects relating to rural communities, including sanitation, water purification, heating, clothing and construction, as well as culturally relevant educational curriculum products are considered.

## Education and Human Resource Development Opportunities

The Arctic Research Program of the OPP is committed to educational and human resource development. Many opportunities exist, primarily through programs in or joint funding with the Education and Human Resources Directorate (EHR). Information about specific programs may be found in the Guide to Programs (NSF 97-150), the NSF Home Page (http://www.nsf.gov), as well as the Elementary, Secondary, and Informal Education (ESIE) Program Announcement (NSF 97-20) and the Undergraduate Education (DUE) Program Announcement (NSF 97-29). Some of the grant and supplement programs supported by the Arctic Research Program are listed below.

## Research Grant Programs

*Knowledge and Distributed Intelligence (KDI)*

KDI is a Foundation-wide effort designed to catalyze the growth in computer power, connectivity, content, and flexibility that is so fundamental that it is dramatically reshaping relationships among people and organizations, and quickly transforming our processes of discovery, learning, exploration, cooperation, and communication. Deadline for KDI submissions is April 1 for Letter of Intent and May 8 for full proposal submission (for details see the KDI Home Page, http://www.nsf.gov/kdi).

*Life and Earth's Environment (LEE)*

LEE is a broad theme describing activities that focus on interdependencies among living organisms and their environment. Emphases may change from year to year, but for FY 99 they include:

- LExEn, (NSF 97-157 or http://www.nsf.gov/pubs/1997/nsf97157/nsf97157.txt) interdisciplinary research program that explores the relationships between microorganisms and the environments within which they exist, with a strong emphasis upon those life-supporting environments that exist near the extremes of planetary conditions. In addition, the Life In Extreme Environments program will explore planetary environments in our own solar system and beyond to help identify possible sites for life elsewhere.
- Environmental Observatories
- Global Change
- Engineered Systems
- Urban Communities
- Integrated Research Challenges

*Doctoral Dissertation Research*

Dissertation grants are available in all OPP disciplines as part of the OPP Arctic Research Program. This support covers travel, fieldwork expenses, data management and other costs connected with doctoral research projects. Proposals are limited to 10 pages and are submitted by the dissertation advisor with the student as co-investigator (Co-PI). Contact the relevant program officer for more information.

*Research Experience for Undergraduates Sites (REUs)*

REU Site Awards provide opportunities for undergraduate students to participate in research projects that support at least six students. The annual submission deadline for the REU Sites Program is September 15.

*Faculty Early Career Development Program (CAREER)*

This program (NSF 97-91) is a Foundation-wide activity that supports junior faculty within the context of overall faculty development. It supports combined research and education activities. Deadline for CAREER submissions is July 22 (for details see the NSF Home Page, http://www.nsf.gov under "cross-cutting programs").

*Increasing Participation of Underrepresented Populations*

NSF supports a number of activities directed at attracting students to science and engineering from underrepresented groups and increasing the numbers of women (e.g., Professional Opportunities for Women in Research and Education, NSF 97-91), and persons with disabilities (NSF 91-54 and NSF 97-85) who are full participants in the mainstream of the Nation's research activities.

## Supplements to Existing OPP Grants

*Research Experiences for Undergraduates (REU)*

Supplements to existing grants in support of one or two undergraduate students in on-going research can be requested at any time. Contact the relevant OPP program officer for more information.

*Informal Science Education (ISE)*

Supplements for Public Understanding of Research. These supplements (of up to $50,000 to existing NSF research grants) are intended to inform the general public about the content, process, and relevance of state-of-the-art research (see NSF 97-70). Interested PIs with active research grants should contact their program officer for information about requirements for these supplements and procedures for applying.

*High School Teacher/Student Arctic Research Experience*

In conjunction with the Education and Human Resources Directorate (EHR) researchers with on-going Arctic projects can volunteer to host a high school teacher/student pair and include them in a research program. Funding is in the form of supplements made to existing grants.

*World Wide Web Supplements*

Small supplements designed to fund the dissemination of Arctic research results to a general audience through the World Wide Web may be available for existing grants. Contact the relevant OPP program officer.

## HOW TO PREPARE PROPOSALS

Before writing a proposal send for a copy of the Foundation's booklet, Grant Proposal Guide (NSF 98-2), or consult the NSF Web page (www.nsf.gov). This booklet gives the format for proposals, lists the budget items that may be supported, explains the proposal evaluation process, and summarizes responsibilities of the grant recipient. Copies of the GPG or the NSF Proposal Forms Kit (NSF 98-3) may be ordered from:

NSF Clearinghouse
PO Box 218
Jessup, MD 20794-0218
Telephone: 301-947-2722
e-mail: pubs@nsf.gov

Contact a program manager in your scientific discipline if you have further questions, especially concerning specific annual program opportunities or consult the NSF Home Page (www.nsf.gov) for new announcements of research opportunities.

### Who May Submit

The National Science Foundation supports researchers affiliated with U.S. universities, research institutions or other organizations, including local or State governments. All applications must be submitted by the sponsoring institution. In accordance with Federal statutes and regulations and NSF policies, no person on grounds of race, color, age, gender, national origin or disability shall be excluded from receiving assistance from the National Science Foundation. The Foundation strongly encourages women, minorities and persons with disabilities to submit proposals to all programs. Arctic research that addresses Native concerns and involves Native collaboration and training is particularly encouraged.

### OPP Target Dates and Proposal Submission

In 1998 the OPP Arctic program will have two target dates for submission of proposals: April 1 and August 1. In subsequent calendar years the target dates will be February 15 and August 1. Proposals for workshops, exploratory research (SGER) or dissertation improvement grants can be submitted at any time.

Proposals for field programs requiring research support in the categories listed on the Logistics Coordination Form must be submitted with sufficient lead time to ensure scheduling and availability. Proposals requesting those logistics capabilities must be submitted no later than the February 15 (April 1 in 1998) target date of the calendar year preceding that in which the research will be conducted. Proposals requiring an oceanographic research vessel must be submitted to either the Division of Ocean Sciences by February 15 of the year preceding the proposed cruise dates or to the OPP Arctic Program by the February 15 (April 1 in 1998) target date to allow 9 months pre-cruise notification. A minimum 9 month advance notification is required for research vessel clearances for Russian waters. Proposals for field work not requiring research support capabilities listed on the Logistics Coordination Form must be submitted no later than the August 1 target date of the preceding year.

## For More Information

For further information about activities mentioned in this announcement, contact the corresponding office listed below, or visit the OPP Home Page (http://www.nsf.gov/od/opp) on the World Wide Web.

The Office of Polar Programs
(703) 306-1030

Arctic Sciences Section
(703) 306-1029

Antarctic Sciences Section
(703) 306-1033

## Other Support

Other NSF programs supporting arctic research have varied deadlines or target dates. Contact specific programs for these dates and refer to the NSF Bulletin for announcements of program deadlines and target dates. The NSF Bulletin and other publications are available through e-mail.

*Proposal Submission*

For hard copy submissions, the original and 20 copies must be received by the target dates. Proposals must be mailed and addressed as follows:

Announcement No_________ or NSF Program_________.
National Science Foundation PPU, Room P60
4201 Wilson Blvd
Arlington VA 22230.

Proposals should be prepared in accordance with instructions in the brochure, Grant Proposal Guide (NSF 98-2). This brochure includes application forms which may be photocopied. It can be obtained from your institution's research office or from the NSF Forms and Publications Office, (703) 306-1130. Proposals may also be submitted electronically using the NSF FastLane system for electronic proposal submission and review, available through the World Wide Web at the FastLane home page (http://www.fastlane.nsf.gov). In order to use NSF FastLane to prepare and submit a proposal you must use a browser that supports multiple buttons and file upload (e.g., Netscape 2.0 and above for Windows, UNIX, or Macintosh). In addition, Adobe Acrobat Reader is needed to view and print forms, and Adobe Acrobat 3.0 (or Adobe Exchange or Distiller) is needed for creating PDF files. To access the FastLane Proposal Preparation application, your institution needs to be a registered FastLane institution. A list of registered institutions and the FastLane registration form are located on the FastLane home page.

Collaborative proposals can be submitted by two or more institutions. These proposals should all be identical, and should contain copies of all budgets, biographical sketches, other support statements, prior support, and of all cover sheets. Each institution should submit its own set. Twenty copies of the lead institution proposal, and five copies of the other institution(s) proposals are needed.

# APPENDIX B

# Biographical Sketches of Committee Members

**John T. Andrews** (*Chair*) is a professor at the University of Colorado at Boulder, Institute of Arctic and Alpine Research. His main research is in the Quaternary history of Arctic areas with special interest in glacial and glacial marine systems. His work has addressed glacial marine processes at high latitudes, reconstruction and history of large Quaternary ice sheets, paleoceanography on high latitude shelves and adjacent seas, and paleoclimatology of arctic lakes and estuaries. He has extensive field experience in the Arctic, having worked in Alaska, Canada, Europe, Greenland, and Iceland as well as in the Ross Sea, Antarctica.

**Susan K. Avery** is director of the Cooperative Institute for Research in Environmental Sciences at the University of Colorado, Boulder. Her research involves the dynamics of the mesosphere and stratosphere, unifying observational analyses and theoretical studies, modeling large-scale atmospheric waves, and ground-based measurement techniques to observe the atmosphere.

**Marianne S.V. Douglas** is an assistant professor in the Department of Geology at the University of Toronto, Ontario. Her research interests are arctic limnology, diatom ecology and taxonomy, paleolimnology, paleoecology, paleoclimates, autecology. She has been involved in multiple limnological surveys of arctic regions.

**Bernard Hallet** is a professor in the Department of Geology and the Quaternary Research Center, University of Washington, Seattle. His research interests are permafrost studies and glacial and periglacial geomorphology, especially processes that shape the landscape in arctic and alpine areas. He is a member of the Polar Research Board.

**Paul A. Mayewski** is director of the Climate Change Research Center and a professor in the Institute for the Study of Earth, Oceans and Space and Department of Earth Sciences at the University of New Hampshire. His expertise is in paleoclimatology, glaciology, and ice core research; and his research interests include climate and environmental change and environmental statistics. He has extensive field experience in the Arctic, Antarctic, and Asia and has been an active leader in a variety of major global change initiatives.

**James H. Morison** is an oceanographer and the department chairman of the Polar Science Center/Applied Physics Laboratory, College of Ocean and Fishery Sciences, University of Washington, Seattle. He developed the Arctic Profiling System for measuring vertical profiles of conductivity, temperature, and velocity, and other scientific equipment. His research interests include seasonal variation and hydrography of the Arctic Ocean and autonomous vehicle and hydrographic buoy measurements and related ocean processes. Dr. Morison is a member of the Polar Research Board.

**Kim M. Peterson** is a professor of biological sciences at the University of Alaska, Anchorage. His research interests are in arctic landscape ecology and climate change effects, especially the carbon balance of tundra and taiga and the effects of increased $CO_2$ and methane, and regional ecosystems integration. He has extensive experience in understanding the structure and functions of arctic ecosystems.

**Donald B. Siniff** is a professor in the Department of Ecology, Evolution, and Behavior at the University of Minnesota. He has broad research interests in ecology and biometry, including vertebrate ecology, statistical and computer applications in field studies, and population dynamics of large mammals. He is experienced in both Arctic and Antarctic research, and at one point served as a commissioner for the U.S. Marine Mammal Commission.

**Roger W. Smith** is associate director of the Geophysical Institute, University of Alaska, Fairbanks. His research interests are in solar-terrestrial physics, including Doppler shift and spectral features in the aurora and airglow, and the dynamics and thermodynamics of the upper atmosphere.

# APPENDIX C

# Guide to Acronyms

| | |
|---|---|
| ANS | Arctic Natural Sciences program |
| ASSP | Arctic Social Sciences program |
| ESH | Earth System History program |
| FY | Fiscal Year |
| MRI | Major Research Instrumentation |
| NRC | National Research Council |
| NSF | National Science Foundation |
| OPP | Office of Polar Programs |
| PI | principal investigator |
| USCGC | U.S. Coast Guard Cutter |